The R Student Companion

The R Student Companion

Brian Dennis

CRC Press
Taylor & Francis Group
Boca Raton   London   New York

CRC Press is an imprint of the
Taylor & Francis Group, an **informa** business

A CHAPMAN & HALL BOOK

CRC Press
Taylor & Francis Group
6000 Broken Sound Parkway NW, Suite 300
Boca Raton, FL 33487-2742

Printed in the United States of America on acid-free paper
Version Date: 20120829

International Standard Book Number: 978-1-4398-7540-7 (Paperback)

Library of Congress Cataloging-in-Publication Data

Dennis, Brian, 1952-
　The R student companion / Brian Dennis.
　pages cm
　Includes bibliographical references and index.
　ISBN 978-1-4398-7540-7 (pbk.)
　1. R (Computer program language) 2. Probabilities. 3. Mathematical statistics--Data processing. I. Title.

　QA276.45.R3D46 2013
　519.50285'5133--dc23

2012025219

Visit the Taylor & Francis Web site at
http://www.taylorandfrancis.com

and the CRC Press Web site at
http://www.crcpress.com

For Chris, Ariel, Scott, and Ellen,

who make me so proud.

Contents

Preface

R is a computer package for scientific graphs and calculations. It is written and maintained by statisticians and scientists for scientists to use in their work. It is easy to use, yet is extraordinarily powerful. R is spreading rapidly throughout the science and technology world, and it is setting the standards for graphical data displays in science publications.

R is free. It is an open-source product that is easy to install on most computers. It is available for Windows, Mac, and Unix/Linux operating systems. One simply downloads and installs it from the R website (http://www.r-project.org/).

This book is for high school and college students, and anyone else who wants to learn to use R. With this book, you can put your computer to work in powerful fashion, in any subject that uses applied mathematics. In particular, physics, life sciences, chemistry, earth science, economics, engineering, and business involve much analysis, modeling, simulation, statistics, and graphing. These quantitative applications become remarkably straightforward and understandable when performed with R. Difficult concepts in mathematics and statistics become clear when illustrated with R.

The book starts from the beginning and assumes the reader has no computer programming background. The mathematical material in the book requires only a moderate amount of high school algebra.

R makes graphing calculators seem awkward and obsolete. Calculators are hard to learn, cumbersome to use for anything but tiny problems, and the graphs are small and have poor resolution. Calculating in R by comparison is intuitive, even fun. Fantastic, publication-quality graphs of data, equations, or both can be produced with little effort. High school and college courses in science and mathematics that currently rely on graphing calculators could benefit greatly from using R instead.

This book will introduce enough R to allow the reader to provide sophisticated solutions to quantitative problems in the sciences and social sciences. But R is huge, and this book is not meant as a comprehensive guide to R. Much of the myth that R has a "steep learning curve" arises from the dizzying complexity of introductory manuals. To become a proficient R user from scratch, one needs a place to begin!

This book will help the reader get started in R. The idea of the book is that one should start by building a basic set of R skills that are learned well. The skill set will handle the quantitative problems arising in most high school and college science courses as well as serve as a base for exploring more advanced material. Part of the fun and rewards of R is discovering for oneself the immense resources available in R.

It is the author's experience that students who have trouble learning R often are actually having more trouble with the underlying mathematical concepts behind the analysis. This book assumes only that the reader has had some high school algebra. Several of the chapters explore concepts from algebra that are highly useful in scientific applications, such as quadratic equations, systems of linear equations, trigonometric functions, and exponential functions. Each chapter provides an instructional review of the algebra concept, followed by a hands-on guide to performing calculations and graphing in R. The chapters describe real-world examples, often drawn from the original scientific publications, and the chapters then show how the scientific results can be reproduced with R. R puts contemporary, cutting-edge quantitative science within reach of high school and college students.

R has a well-deserved reputation as a leading software product for statistical analysis. However, R goes way beyond statistics. It is a comprehensive software package for scientific computations of all sorts, with many high-level mathematical, graphical, and simulation tools built in. Although the book covers some basic statistical methods, it focuses on the broader aspects of R as an all-round scientific calculation and graphing tool.

Another part of the mythical difficulty of learning R stems from the problem that many of the books and web sites currently available about R are also about statistics and data analysis. This book, however, is prestatistics and largely avoids the prepackaged statistical routines available in R. The concepts of statistical inference are challenging. The book introduces some computational probability and simulation, some summary statistics and data graphs, and some curve fitting, but the book does not dive into statistical inference concepts. In fact, students who have the R background contained in this book are positioned to get far more out of their initial exposure to statistical inference.

This book does *not* assume that the reader has had statistics or calculus. Relatively few students take calculus before college, and even fewer take statistics before their middle years in college. Instead, this book concentrates on the many uses of R in precalculus, prestatistics courses in sciences and mathematics. Anything in science, mathematics, and other quantitative courses for which a calculator is used is better performed in R. Moreover, R greatly expands the complexity of the scientific examples that can be tackled by students.

Students who use R in their science courses reap great benefits. With R, scientific calculations and graphs are fun and easy to produce. A student using R is freed to focus on the scientific and mathematical concepts without having to pore through a manual of daunting lists of calculator keystroke instructions. Those calculator instructions are never actually mastered and internalized, and they change with each new machine. R skills by contrast are mastered and grow with each use, and they follow the student on into more advanced courses. The students will be analyzing data and depicting equations just as scientists are doing in laboratories all over the world.

R invites collaboration. Like the scientists sharing their applications, students can work in groups to conduct projects in R, build R scripts, and improve each others' work, and they can collect, accumulate, and just plain show off exemplary graphical analyses. Results on a computer screen are much easier to view in groups than on a calculator, and R scripts are much easier to save and alter cooperatively than are the calculator keystroke lists. At home, students can message each other about their latest R scripts, working cooperatively online. Every new class can take what previous classes have done and build new accomplishments upon the old. *R builds on itself.*

R use has exploded in colleges and universities, and R has recently been adopted by forward-looking technology companies (see "Data Analysts Captivated by R's Power," by Ashlee Vance, *The New York Times*, January 6, 2009). A simple online search will reveal the extent to which R is shaping education in advanced university courses. Not only do R skills follow a student throughout their coursework, but knowledge of R is also a bona fide professional scientific credential of substantial value.

Online web resources for R use are vast. Scientists and professors have leaped at the opportunity to share their accumulating knowledge about using R in scientific applications. Free web-based tutorials, primers, and reference books are posted on scientists' and courses' web pages everywhere. Discussion forums about R exist, where people can get questions answered and share R "scripts" for reproducing the calculations and graphics from the latest scientific paper. These resources, however, tend to be targeted at more advanced and specialized college level courses. For the beginner, the sheer preponderance of material makes it hard to know where to start. Most R users have had to struggle initially with the need to cope with and filter the disorganized cloud of R instructional material.

This book is designed with the beginner in mind. With this book, the student can master an initial set of R techniques that will handle most computation and graphing projects in basic science and applied mathematics courses. The techniques are here in one place, easy to learn and easy to find. The student can then graduate to the more advanced techniques with confidence.

How to use this book: This book is meant to be *done*, not read! First, install R on your computer or locate an institutional computer on which R has been installed. Then, start at the beginning of the book and follow along and type the R commands as you go. Save your work in a folder on the computer hard drive or on portable storage media. At the end of each chapter, tackle one or more of the computational challenges. These are more than just exercises; they are miniprojects that will help you build R-thinking into your creativity and problem-solving. In a course setting, the computational challenges can be parceled out among students or groups of students, and their results can be shared with or reported to the class as a whole. The appendices at the end of the book provide additional advice about installation of R, information about getting help with R techniques, and a handy listing of the most commonly used R commands and options.

All the scripts presented in this book, all the scripts which produced each figure in the book, and all the data sets in this book are posted at http://webpages.uidaho.edu/~brian/rsc/RStudentCompanion.html.

Readin', Ritin', Rithmetic … and R!

Brian Dennis
Moscow, Idaho

Author

Brian Dennis is a professor with a joint appointment in the Department of Fish and Wildlife Sciences and the Department of Statistical Sciences at the University of Idaho. He received his master's degree in statistics and his PhD in ecology from the Pennsylvania State University. He has authored more than 70 scientific articles on applications of statistics and mathematical modeling in ecology and natural resource management. He has enthusiastically used R in his work and teaching R in his courses for a decade.

1

Introduction: Getting Started with R

R is a computer program for doing scientific graphs and calculations.

R was written by scientists, for scientists to use in their work.

R is incredibly powerful and amazingly easy to use.

R is free.

Did I mention that R is free?

Versions of R are available for Windows computers, Macs, and even Unix or Linux. If you have a computer at home, you can download and install R from the web site http://www.r-project.org/.

Installation is easier than installing a computer game (but if you need help, refer to Appendix A).

Installing R will put an icon, a blue "R," on the computer desktop or in the program list. Find it and click it to start the program. The R program window will appear, and the "R console," a window within the R program window, will pop up.

On the console, you will see the prompt ">" followed by a cursor. Now, R is waiting for your instructions! Just type the commands that appear at the prompt as you go through this tutorial, or follow along as your instructor demonstrates the commands on a projector. Afterward, work alone or in groups to answer the computational challenges at the end of this chapter. Be prepared to give a short presentation of your results to the class. Ready? Let's get started.

R Tutorial

The simplest way of using R is as a powerful calculator. At the prompt, type: 5+7 and hit the Enter key:

```
> 5+7
[1] 12
```

Part 1 of the answer is 12. We will later see that answers can often have many parts, and so R prints part numbers along with the answers.

Let us try subtraction. Each time, type the characters after the prompt and hit the Enter key; the answer you should see is then printed in this tutorial on the next line:

```
> 5-7
[1] -2
```

R knows about negative numbers. Let us try:

```
> 5+-2
[1] 3
```

Now, multiplication is represented by an asterisk "*":

```
> 5*7
[1] 35
```

Division is indicated by a forward slash "/", so 5 divided by 7 is

```
> 5/7
[1] 0.7142857
```

R will do calculations as decimal numbers, just like a calculator.

Raising to a power: Recall that "five to the seven power," written as 5^7, means $5 \times 5 \times 5 \times 5 \times 5 \times 5 \times 5$. R will calculate that for us. The caret symbol "^" denotes raising to a power. Thus, five to the seven power is

```
> 5^7
[1] 78125
```

You can put a string of calculations all in one command. The calculations with * and / are done first, then with + and –, and the calculations are done from left to right:

```
> 5+7*3-12/4-6
[1] 17
```

See if you can get the above answer by hand. Also, raising to a power is done first, even before multiplication and division:

```
> 1+4*3^2
[1] 37
```

You can alter the priority of operations by using parentheses:

```
> (5+7)*3-(12/4-6)
[1] 39
```

Parentheses inside of parentheses will be done first! Just ensure that every left parenthesis "(" has a right friend ")":

```
> (5+7)*3-(12/(4-6))
[1] 42
```

R will store *everything* for you. You just give names to your calculations:

```
> sally=5+7
> ralph=4-2
> sally-ralph
[1] 10
```

When giving names, remember that R notices lowercase and uppercase characters. In R, the name sally is different from Sally.

Are ralph and sally still there?

```
> sally
[1] 12
> ralph
[1] 2
```

They will disappear when you exit the R program without saving the "workspace."

If you use the name for something different, R will erase the old value:

```
> ralph=9
> ralph
[1] 9
```

Interestingly, the symbol "=" in R (and in many computer programming languages) does not mean "equals." Instead, it means

calculate what is on the right, and store the result using the name on the left.

This was designed by computer scientists seemingly to irritate their former math teachers. For example, ralph=ralph+1 is a statement that mathematicians do not like, because no number exists that, when you add one to it, you get the same number! However, the statement makes perfect sense to computer programmers. It means

take the old value of ralph, add 1, and store the result as the new value of ralph.

A statement with an equals sign is called an "assignment statement," assigning the value resulting from the calculation on the right to the storage location on the left. Try it:

```
> ralph=ralph+1
> ralph
[1] 10
```

Actually, previous versions of R used the syntax ralph<-ralph+1 for assignment statements (the word "syntax" means typographic rule). The symbol "<-" (a "less than" symbol followed by a hyphen) is supposed to look like a little arrow pointing left. Many older web sites and books about R use this syntax, and in fact the syntax still works in the current versions of R. Let us try it:

```
> sally<-sally+ralph
> sally
[1] 22
```

The assignment statement calculated and stored 22 as the new value of `sally`. The scientists responsible for R finally gave up trying to be mathematical purists and instituted the equals sign for assignment statements in order to be consistent with most other computer languages.

Now, we are ready to unleash some of the power of R!

Vectors

R can work with whole "lists" of numbers. Try this:

```
> x=c(3,-2,4,7,5,-1,0)
> y=4
> x+y
[1]   7   2   8  11   9   3   4
```

The `c()` command in the first line above says "combine" the numbers 3, –2, 4, 7, 5, –1, and 0 into a list. We named the list x. R has a special term for a list of numbers: a **vector**. Here, x is a vector with seven **elements**. The value of y is 4. The expression x+y added 4 to every value in x! But what if y were a vector like x?

```
> y=c(1,2,3,4,5,6,7)
> z=x+y
> z
[1]   4   0   7  11  10   5   7
```

Each number in the vector y is added to the corresponding number in x!

Remember in fourth grade when your teacher gave you a bunch of big multiplications to do for homework:

75,634	2,339	103,458	48,761	628,003
× 567	× 138	× 974	× 876	× 402

Put the top numbers in a vector (let us name it "top"), and the bottom numbers in another vector (say, named "bot"). Then, multiply the vectors:

```
> top=c(75634,2339,103458,48761,628003)
> bot=c(567,138,974,856,402)
> top*bot
[1]   42884478    322782    100768092   41739416   252457206
```

There are a few things to note here. (1) When writing R statements, do not use commas within large numbers to group the digits in threes. Rather, commas are used in R for other things, such as to separate numbers in the c() (combine) command. (2) Enter the numbers in the two vectors in the same order. (3) Spaces in between the numbers are fine, as long as the commas are there to separate them. (4) Do not show this to your younger sibling in fourth grade.

All of the arithmetic operations, addition, subtraction, multiplication, division, and even power, can be done in R with vectors. We have seen that if you operate with a single number and a vector, then the single number operates on each element in the vector. If you operate with two vectors of the same length, then every element of the first vector operates on the corresponding element of the second vector.

The priority of operations is the same for vector arithmetic, and parentheses may be used in the usual way to indicate which calculations to perform first:

```
> ted=c(1,2,3)
> kat=c(-1,1,.5)
> 2*(ted+kat)
[1]   0   6   7
> 2*ted+kat
[1]   1   5   6.5
```

If you make a mistake while typing, just type the line again. R will calculate and store the newer version. Also, if a line is long, you can continue it on the next line by hitting the Enter key *at a place where the R command is obviously incomplete* (R is remarkably smart!). R will respond with a different prompt that looks like a plus sign; just continue the R command at the new prompt and hit the Enter key when the command is complete:

```
> kat=c(-1,1,
+ .5)
> kat
[1]  -1.0   1.0   0.5
```

A special vector can be built with a colon ":" in the following way:

```
> j=0:10
> j
 [1]   0   1   2   3   4   5   6   7   8   9  10
```

Here, j was defined as a vector consisting of all the integers from 0 to 10. One can go backward if one wants:

```
> k=5:-5
> k
 [1]  5  4  3  2  1  0 -1 -2 -3 -4 -5
```

Do you want to see the powers of 2 from 2^0 to 2^{20}? Of course you do:

```
> j=0:20
> 2^j
 [1]       1       2       4       8      16      32      64     128
 [9]     256     512    1024    2048    4096    8192   16384   32768
[17]   65536  131072  262144  524288 1048576
```

You should note here that the text syntax in R for writing math expressions forms a completely unambiguous way of communicating about math homework problems via instant messaging or text messaging:

> *Ralph:* hey, what's up?
>
> *Sally:* working on math homework. Yuck.
>
> *Ralph:* oh darn, yeah, and I 4got to write down all that "solving a quadratic" stuff.
>
> *Sally:* here it is...
>
> *Sally:* a*x^2 + b*x + c = 0.
>
> *Sally:* there are two solutions provided b^2 > 4*a*c.
>
> *Sally:* (-b+sqrt(b^2-4*a*c))/(2*a).
>
> *Sally:* (-b-sqrt(b^2-4*a*c))/(2*a).
>
> *Ralph:* thx! R U using that R thing to do the homework?
>
> *Sally:* of course! Using a calculator would take twice as long.
>
> *Ralph:* want to meet at the coffee shop after ur done?
>
> *Sally:* sure! I am almost finished, thx to R.

Sally and Ralph are experienced R users and know that sqrt() takes the square root of whatever is inside the parentheses. We will look at that and other functions in Chapter 3.

Graphs

Are you ready for a graph? If you are not impressed yet by R, prepare to be so. Suppose you have accumulated $1000 and would like to save it for future use, perhaps for buying a house. Suppose you find a bank that offers a certificate of deposit (CD) paying interest of 5% per year, which will be reinvested in the CD. Such a CD would be a great opportunity to put your money to work for you.

Let us see why, by drawing a graph in R. The graph will show the amount of money in the CD after year 1, year 2, and so on, up to, say, year 10.

For interest of 5% compounded annually, we multiply the amount of money in the CD each year by $(1+0.05)$ in order to calculate how much money is in the CD at the end of the following year. So, we calculate the amount of money after year 1 by $1000(1+0.05)$. We calculate the year 2 amount by $1000(1+0.05)(1+0.05)$, the year 3 amount by $1000(1+0.05)(1+0.05)(1+0.05)$, and so on. See the pattern? We can represent the money after year t as an equation, based on the pattern. If n is the number of dollars in the CD after year t, then the equation is

$$n = 1000(1+0.05)^t.$$

Drawing a picture of the changing number of dollars through time according to this equation is a ready-made task for R. (1) We will set up a vector called "t" that will contain the years 0, 1, 2, 3, ..., 10. (2) We will calculate a vector called "n" using the equation; n will contain all the dollar amounts in the CD corresponding to the years 0, 1, 2, 3, ..., 10. (3) We will draw an $x - y$ graph of the values of n (vertical axis) versus the values of t (horizontal axis), connecting the points with lines. Type in the following R commands (be sure to note that "l" in the third command is a lowercase "L," not the numeral "one"):

```
> t=0:10
> n=1000*(1+0.05)^t
> plot(t,n,type="l")
```

A separate graph window should pop up on your screen as shown in Figure 1.1. We used only three commands! Pretty cool, huh? The plot() command is a built-in routine in R for two-dimensional plots. There are many such graphical routines in R for many different kinds of graphical displays. Most of them can be customized to accommodate virtually anything that one might want to include in a display, such as different axis labels, tic marks, titles, symbols, and symbol legends.

In the plot() command, the horizontal axis variable is listed first, followed by the vertical axis variable and various optional statements setting the appearance of the graph. Here, the type="l" option (l stands for "line") specifies the type of plot to be a line plot with points connected by lines, but no symbols drawn for the points themselves. R automatically picks appropriate axis sizes, but they can be altered with additional options in the plot statement, along with axis labels, tic marks, title, line thicknesses, and so on. Appendix C lists the different types of plots available, and you will use many of them as you work through the chapters in this book. The entries or "arguments" in the plot() command are separated by commas, the standard syntax for the arguments and options entered in built-in routines in R.

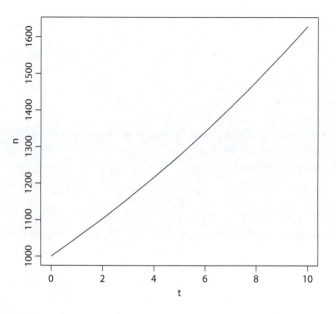

FIGURE 1.1
Amount of dollars *n* after year *t* in a certificate of deposit paying 5% interest compounded annually, with an initial investment of $1000.00.

The graph is a graphical object that can be saved as a file in various graphic formats. Click on the graph window to make it active, then look in "File" on the top menu bar, and select "Save As." Good graphical formats for such scientific plots are EPS or PDF. From the File menu, you can alternatively copy the graph to the clipboard and subsequently paste the graph into a word processor document.

The graphical object is actually "open" in R, waiting for you to add further points, curves, annotations, and so on. You will learn many such customizations in subsequent chapters.

When you are done with the graph and are ready to make another, close the graph window.

Real-World Example

Although it is hard to exceed the real-world importance of money, we chose the example of CD investment mentioned earlier mainly for its simplicity. Many chapters of this book will close by tackling more complex, real-world examples that will require putting R calculations and graphics together in a cumulative and comprehensive way.

Let us try a graph of some data from ecology. This is an example from real science, not a "toy" example, and so it requires a bit of explaining.

Ecology is a subfield of biology that deals with the study of the relationships between organisms and their environments. Predators and their prey is a topic that has excited ecologists for many decades, and the topic is important to society. For instance, questions about wolf predation on deer, elk, moose, and livestock have become politically controversial in some parts of the United States. The reintroductions of wolves in locales where they once were exterminated as pests have the potential to adversely affect hunting and livestock production.

How many moose do wolves eat in the wild? Well, we might expect that the average number of moose killed by an average wolf would depend on the supply of moose! Getting some idea about the form of the relationship between wolf feeding rate and moose supply would be helpful to wildlife managers who must make decisions about wolf culling and moose hunting over a wide range of moose and wolf abundances.

In the table below are some figures that an ecologist assembled together from studies across wolf ranges in North America (Messier 1994). In the regions studied, moose were the preferred food of the wolves and occurred in varying densities. Wolves live and hunt mostly in packs, and the moose kill rates are calculated as the average number of moose killed per wolf per 100 days:

moose density	kill rate
0.17	0.37
0.23	0.47
0.23	1.90
0.26	2.04
0.37	1.12
0.42	1.74
0.66	2.78
0.80	1.85
1.11	1.88
1.30	1.96
1.37	1.80
1.41	2.44
1.73	2.81
2.49	3.75

Here, "moose density" is the average number of moose per 1000 km^2. One thousand square kilometers is an area roughly equivalent to a square of 20 miles by 20 miles. A glance at the numbers suggests that moose—and wolves—use a lot of land: one moose or so per thousand square kilometers means that wolves must range far and wide for a square meal! But the numbers by themselves are just a stark table and do not convey much more to us beyond their range of values. Instead, a visual portrayal of the numbers will help reveal any relationships that might exist.

To explore these data, we will simply plot each pair of numbers (moose density and kill rate) as a point on an *x–y* graph. Scientists call that type of graph a scatterplot. We need two vectors containing the values to be plotted. Let us call them "moose.density" and "kill.rate." Scientists find it helpful to give descriptive names to the quantities and objects in R calculations, and when choosing names, they will often string together several words connected by periods. After setting up the vectors (and checking the numbers carefully), we then just add the plot() command, only this time using the type="p" option:

```
> moose.density=c(.17,.23,.23,.26,.37,.42,.66,.80,1.11,1.30,1.37,
+ 1.41,1.73,2.49)
> kill.rate=c(.37,.47,1.90,2.04,1.12,1.74,2.78,1.85,1.88,1.96,
+ 1.80,2.44,2.81,3.75)
> plot(moose.density,kill.rate,type="p")
```

You should now see a scatterplot of the points! The type="p" option in the plot() command produces the scatterplot (p stands for "points"), with symbols drawn for the points without connecting the points with lines. The first two commands for putting the data into the vectors used continuation lines ("+" prompt) in order to fit the commands compactly within the typesetting of this book, but most R consoles will accept very long commands in one line.

Let us add something to our graph! How about a mathematical curve summarizing ecologists' current understanding about what the relationship should look like? We will then have scientific hypothesis and real-world data compared on one graph.

Before adding such a curve, we will try to make some sense about what we see on the graph. The data are scattered, "noisy" as scientists say, but there is a sort of pattern. As moose density increases, so does the kill rate, but then the kill rate seems to flatten and does not continue increasing as fast.

Ecologists have seen this pattern in many predator–prey systems, and the current hypothesis for the cause of the pattern goes something like this. If moose are rare, a given wolf will likely consume very few moose in a fixed period of time, so the data point will be near zero for both axes. If the supply of moose increases, we can hypothesize that the number of moose consumed per wolf would also increase.

But what if moose were very abundant, as if nature staged an "all-you-can-eat" banquet for wolves? We would not expect the average wolf's rate of consumption of moose to increase without bound, because the physical capacity for killing, handling, and digesting moose is limited. Instead, we might expect that as the supply of moose increases from abundant to very abundant, the number of moose consumed by an average wolf in the given period of time would simply level off. There is an upper physical limit to the speed with which wolves can hunt and eat moose, just as there is an upper limit to the speed with which you can eat hamburgers (file that under biology lab exercises you would love to try).

To summarize, we conjecture that the relationship between the feeding rate of an average wolf and the supply of moose, if plotted on an x–y graph (with moose supply as the variable on the horizontal axis) would start near zero, rise steeply at first, but then gradually turn toward horizontal at some upper maximum feeding rate.

Ecologists have expressed this hypothesis in a mathematical model, an equation that captures the essential expected relationship under "ideal" circumstances, that is, only the moose supply is being varied, with any other environmental variables that could affect the situation being held fixed. The equation is as follows:

$$k = \frac{am}{b+m},$$

where k is the kill rate of an average predator, m is the supply of prey, and a and b are constants that have different values for each type of predator and each type of prey (lady beetles eating aphids have different values for a and b than do wolves eating moose). For the wolf–moose data, the scientist obtained the following values for a and b by using a sophisticated "curve-fitting" program (like the one you will learn in Chapter 15):

$$a = 3.37,$$

$$b = 0.47.$$

Let us add a plot of the equation to our graph! For this, we need two more vectors. One will contain a range of moose densities (let us call it m). The other will contain the resulting kill rates calculated using the equation (we will call it k). We also need to define the quantities a and b before the equation can be computed. To obtain a nice smooth curve, we should use many values of moose density (a hundred or so), ranging from 0 to, say, 2.5 (the upper end of the moose axis on our graph). The expression (0:100)/100 will produce a vector with a low value of 0 and a high value of 1, with many values in between (check this in the console!). We just multiply that vector by 2.5 to get the vector with a range of values of moose supply from 0 to 2.5. Leave the scatterplot "open" (do not close the graph window), and click on the R console window to make it active. Carefully type the following:

```
> m=2.5*(0:100)/100
> a=3.37
> b=0.47
> k=b*m/(a+m)
```

These commands build the two vectors, m and k, to add to our graph. To put them on our graph in the form of a curve, just type one more statement:

```
> points(m,k,type="p")
```

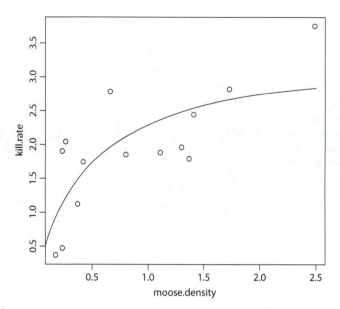

FIGURE 1.2
Data and model depicting the relationship between the wolf rate of feeding on moose and the density of moose. Circles: number of moose killed per wolf per 100 days (vertical axis) with the number of moose per 1000 km² (horizontal axis). Solid line: the "functional response" equation from predator–prey theory (Gotelli 2008) fitted to the data. (Data from Messier, F., *Ecology*, 75, 478–488, 1994.)

The points() command adds extra points onto an open graph. Its arguments and options work mostly like those in the plot() command.

The graph you should be seeing in your R window appears in Figure 1.2. You have essentially reproduced a figure in the original scientific journal article! Save the graph, and save the R commands (How might you do this? Try some ideas!) for future reference.

Final Remarks

It might seem that typing more than a few lines in the command console for a longer, more complex calculation or a complicated graph could become unwieldy. That certainly is true. In Chapter 2, you will find out how to enter, edit, and save a long list of commands into a special R file called a "script," and run the entire list all at once. It might also seem that typing a large data set into one or more vectors using the c() (combine) command is awkward and inconvenient. In Chapter 5, you will find out how to enter and store data in a separate data file and then how to bring the data file into R for plotting and analysis. Be assured that the writers of R know that most people, especially scientists, hate unnecessary work.

WHAT WE LEARNED

1. Priority of operations in arithmetic statements in R: raising to a power (^), multiplication and division (* and /), addition and subtraction (+ and -), operations performed from left to right, and priorities overridden by parentheses.

Example:

```
> 3*(12-5)^2+4-6/2*2
[1] 145
```

2. Store the results in computer memory using assignment statements. Assignment statements use "=" or "<-" symbols, calculating what is on the right of the symbol and storing the result under the name on the left. Names are case sensitive. Type the name to retrieve the quantity.

Example:

```
> Daily.run=2
> Cumulative.run=20
> Cumulative.run=Cumulative.run+Daily.run
> Cumulative.run
[1] 22
```

3. Vectors are ordered lists of numbers. The combine command c() is a handy way of creating small vectors. Arithmetic operations between a single number and a vector are performed using the single number and each element of the vector, using the same priority of operations as for ordinary arithmetic statements. Arithmetic operations for two vectors of the same size are performed on the corresponding elements of the two vectors, using the same priority of operations as for ordinary arithmetic statements. The colon command : provides a handy way of creating a vector containing a sequence of numbers.

Example:

```
> time.long=c(0,1,2,3,4,5,6,7,8,9,10);
> time.long
[1]  0  1  2  3  4  5  6  7  8  9 10
> time.quick=0:10
> time.quick
[1]  0  1  2  3  4  5  6  7  8  9 10
> time.long+time.quick
[1]  0  2  4  6  8 10 12 14 16 18 20
```

```
> distance=16*time.long^2
> distance
 [1]    0   16   64  144  256  400  576  784 1024 1296 1600
```

4. The `plot()` command produces *x–y* plots of two vectors. The horizontal axis variable is listed first in the command arguments. Command arguments are separated by commas. Different types of plot styles are obtained with the `type=` argument, with simple line plots (points connected by lines) designated by `type="l"` and scatterplot (points plotted with symbols but not connected) designated by `type="p"`. Plots can be saved as files under various graphical formats or copied to the clipboard. The `points()` command adds more points to an open plot and uses the same syntax as the original `plot()` command.

Example:

```
> time=(0:10)/10
> distance=16*time^2
> plot(time,distance,type="l")
> lab.data=c(.3,1.4,1.5,1.9,4.8,5.3,7.9,9.8,13.7,15.7)
> lab.time=(1:10)/10
> points(lab.time,lab.data,type="p")
```

Computational Challenges

The computational problems that follow can be accomplished with the R techniques you have learned so far. Expect some snags in your early attempts. You will get the hang of R with practice, trial and error, and consultation with classmates. Remember, if you mistype a command, just type it again, and R will overwrite the previous value. Copy and save your successful commands and results into a word processor file for future reference.

1.1. Evaluate the following expressions:

$$\frac{93^2 - 164}{46^3 + 189} \qquad 376 - \frac{23^2}{4} \qquad \frac{59 + 48^2}{-9 + 22^2} - \frac{-16 + 55^2}{13 + 29^2} \qquad 18^4 - 16^3 + 14^2 - 12$$

$$3^x \text{ for } x = 1, 2, ..., 20 \qquad 5^x \text{ for } x = 1, 2, ..., 10$$

1.2. Draw the investment equation (money in the CD) again, except use a much longer time horizon. Allow time to range as many years into the future (50?) as, say, you have until you reach the retirement age of 65. Gaze at the graph with awe. Yes, it is really your choice: having those fancy designer jeans now or having many times the sale price later!

Add two or three more curves onto your graph showing the effects of two or three different interest rates.

1.3. The following are the population sizes of the United States through its entire history, according to the U.S. Census. Construct a line plot (type="l") of the U.S. population (vertical axis) versus time (horizontal axis). By the way, rounding the population sizes to the nearest 100,000 or so will hardly affect the appearance of the graph.

1790	3,929,214
1800	5,308,483
1810	7,239,881
1820	9,638,453
1830	12,860,702
1840	17,063,353
1850	23,191,876
1860	31,443,321
1870	38,558,371
1880	50,189,209
1890	62,979,766
1900	76,212,168
1910	92,228,496
1920	106,021,537
1930	123,202,624
1940	142,164,569
1950	161,325,798
1960	189,323,175
1970	213,302,031
1980	236,542,199
1990	258,709,873
2000	291,421,906

Repeat the plot six more times (saving the graph each time), using type="p", type="b", type="c", type="o", type="h", and type="l". Compare the different graph types. What different aspects of the data are emphasized by the different graph types?

1.4. If you throw a baseball at an angle of 45°, at an initial velocity of 75 mph, while standing on a level field, the ball's horizontal distance x traveled after t seconds is described (neglecting air resistance) by the following equation from Newtonian physics:

$$x = 27.12t.$$

Furthermore, the height above the ground after t seconds, assuming the ball was initially released at a height of 5 ft, is described by

$$y = 1.524 + 19.71t - 4.905t^2.$$

The equations have been calibrated to give the distance x and height y in meters. The ball will hit the ground after about 4.09 seconds. Calculate a vector (say, x) of baseball distances for a range of values of t from 0 to 4.09. Calculate a vector of baseball heights (say, y) for the same collection of times. Make a plot of x (horizontal axis) and y (vertical axis). Read from the graph of the ball's trajectory how high and how far, approximately, the ball will travel.

NOTE: For different initial throwing velocities and angles, the above baseball equations will have different numerical coefficients in them. The equations can be written in more general form to accept different initial conditions, but to do that, we need a little bit of trigonometry (Chapter 9).

1.5. According to Newton's universal law of gravitation, the acceleration of an object in the direction of the sun due to the sun's gravity can be written in the form

$$a = \frac{1}{r^2},$$

where r is the distance of the object from the sun's center, in astronomical units (AU) of distance. One AU is the average distance of the Earth from the sun, about 150 million kilometers. The units of a are scaled for convenience in this version of Newton's equation so that one unit of acceleration is experienced at a distance of 1 AU. Use the equation to calculate the gravitational accelerations at each of the planets' average distances from the sun:

Planet	Mercury	Venus	Mars	Jupiter	Saturn	Uranus	Neptune	(Pluto)
Distance	0.39	0.723	1.524	5.203	9.539	19.18	30.06	39.53

Pluto is now considered to be a large comet-like object or dwarf planet that originated from the Kuiper Belt.

1.6. Using the equation from Question 1.5, draw a graph of the gravitational acceleration a (vertical axis) versus a range of values of r ranging from around 0.4 AU (distance of Mercury) to around 5.2 AU (distance of Jupiter). According to Newton's gravitational law, is there any distance at which the sun's gravity is escaped entirely?

1.7. Using the kill rate equation for the wolf–moose system that you studied in the tutorial, do some numerical exploration/experimentation to find out where the curve levels off (the maximum kill rate of the average wolf). How is it possible that some of the real data had kill rate values above that maximum? Does the leveling off point resemble any numerical quantity you see in the equation itself?

1.8. To decrease the use of insecticides in agriculture, predator insects are often released to combat insect pests. Coccinellids (lady beetles), in particular, have a voracious appetite for aphids. In a recent study (Pervez and Omkar 2005), entomologists looked at the suitability of using coccinellids to control a particular aphid, *Myzus persicae* (common name is the "green peach aphid"), a serious pest of many fruit and vegetable crops. In the study, the entomologists experimentally ascertained aphid kill rates for three different species of coccinellids:

APHID DENSITY (#/jar)	COCCINELLID FEEDING RATE (# eaten per 24 hr)		
	Cheilomenes sexmaculata	*Coccinella transversalis*	*Propylea dissecta*
25	21	21	15
50	37	37	26
100	65	60	42
200	102	90	59
300	125	109	69
400	141	120	74
500	154	129	79
600	164	136	82
700	170	140	83
800	177	143	85

Enter the data columns above into vectors, giving them descriptive names. For each type of coccinellid, use R to construct a scatterplot (type="p") of the feeding rate of the coccinellid versus aphid density. Then, add a kill rate curve to the coccinelid/aphid graph. Use the following constants in the kill rate equations:

$$C.\ sexmaculata: a = 234.5,\ b = 261.9,$$
$$C.\ transversalis: a = 178.9,\ b = 194.9,$$
$$P.\ dissecta: a = 100.4,\ b = 139.8.$$

Save and close each graph before starting the graph for the next coccinellid.

1.9. Plot the moose–wolf data again in a scatterplot, and save the graph under all the different graphical file formats (.JPG, .EPS, .PNG, etc.) available. Import each version into a word processor or presentation program so that you can compare the graphical formats side by side. Do some Internet research and find out the main advantages and disadvantages of each of the available formats. List these advantages and disadvantages in your document or presentation, in conjunction with the example you made of a scatterplot in each format. Share your discoveries!

References

Gotelli, N. J. 2008. *A Primer of Ecology*. Sunderland, MA: Sinauer.

Messier, F. 1994. Ungulate population models with predation: A case study with the North American moose. *Ecology* 75:478–488.

Pervez, A., and Omkar, A. 2005. Functional responses of coccinellid predators: An illustration of a logistic approach. *Journal of Insect Science* 5:1–6.

2

R Scripts

Even small calculation tasks can be frustrating to perform when they are typed line by line in the console window. If you make a mistake, an entire line or more might have to be retyped. For instance, in Chapter 1, we used three R commands to produce a graph:

```
> moose.density=c(.17,.23,.23,.26,.37,.42,.66,.80,1.11,1.30,1.37,
+ 1.41,1.73,2.49)
> kill.rate=c(.37,.47,1.90,2.04,1.12,1.74,2.78,1.85,1.88,1.96,
+ 1.80,2.44,2.81,3.75)
> plot(moose.density,kill.rate,type="p")
```

If the data in one of the vectors were typed incorrectly, the graph would come out wrong, and the offending statements would have to be reentered. Fun then starts to resemble work.

Happily, R has a feature for easily managing long lists of commands. The feature is called an R script. An R script is a prepared list of R commands that are processed sequentially by R from top to bottom. The R script can be typed, edited, saved as a file, processed in part or in whole, changed, reprocessed, and resaved.

Creating and Saving an R Script

Let us put the wolf–moose graph commands into an R script. Start R on your computer so that you have the R console window ready to go. Now, move the cursor up to the File statement on the task bar and click to get a pull-down menu. Click on New script. A new window, the R editor, should appear. The window is a basic text editor in R for creating R scripts. Note: The R editor exists only in Windows and Mac versions of R. If you are using a computer with the Unix or Linux operating systems, then the current version of R for those systems has no built-in editor for scripts. You must create your scripts in a separate text editor of your choice and then copy/paste the commands into the R console. Many Windows and Mac users of R actually prefer to use their own favorite text editors instead of the one provided by R.

Type the wolf–moose commands one by one into the R editor window. If
you have saved the commands in a file somewhere, you can copy and paste
them into the R editor. Do not include an R-prompt character (">") in any
command in the editor. Hit the Enter key after each command to start a
new line, just like a normal text editor:

```
moose.density=c(.17,.23,.23,.26,.37,.42,.66,.80,1.11,1.30,1.37,
1.41,1.73,2.49)
kill.rate=c(.37,.47,1.90,2.04,1.12,1.74,2.78,1.85,1.88,1.96,
1.80,2.44,2.81,3.75)
plot(moose.density,kill.rate,type="p")
```

All the usual editing tools are available in the R editor window: backspace,
delete, cut, and paste. If you want to continue a command on the next line,
just break it at a place where the command is obviously incomplete (as was
done above for the moose.density and kill.rate statements). The commands are not processed by R yet. You can type the commands, edit them,
and get them ready for R. Check your typing when you are finished. If some
errors escape your attention, they will be easy to fix later without having to
retype the whole script.

Take a moment now to see how R handles windows. Click on the console
window (any part that is visible). The console will become the active window. Here, you can do subsidiary calculations while preparing a script as
well as check on how things created in the script turned out after the script
was processed. Click back on the R editor window to make it active again.

Are you ready to have R "run" these commands? No, not quite yet. You
should save this script first as a file in your computer or on suitable storage
media. To save the script, the R editor must be the active window, and everything in the editor will be saved in the file you designate. On the task bar above,
click on File and then on Save as... to get a file-saving directory window for
your system. Eventually, you will want folders for each of your math and science courses for computational projects, but for now, find a convenient folder
on your system or create a new folder, perhaps named "R projects." Choose
and write a name (perhaps "wolf moose graph") for the script in the file name
box. R does not automatically add any extension (".txt", ".doc", etc.) to the file
name; a good extension to use is ".R". So, "wolf moose graph.R" is now a file
stored in your system. You can access it later as well as share it with coworkers.

Running an R Script

Now you are ready to run the script! With the R editor window active, click
Edit on the task bar. The resulting menu has the usual text editing choices,
Undo, Cut, Copy, and so on. Find and click on Run all.

If you are using R in Unix/Linux, there are two ways to run your script. (1) You can copy the script in your text editor and then paste it in its entirety at the console prompt. The script lines will be entered at the console and executed, just as if you were typing them one by one in the console. (2) You can save your script in a folder on your computer and then use the `source()` command at the console. In the parentheses of the source command, you would enter the directory location of your script file, for example,

```
> source("c:/my_R_scripts/wolf moose graphs.R")
```

Many experienced R users prefer the `source()` method even if they are working in Windows or Mac operating systems. Whatever operating system you are using, R will want the directory in the `source()` command written with forward-slanting slashes.

If the script is free of errors, the graph of the wolf feeding rates and the moose densities will pop up. Things now are exactly as if the three commands had been entered in the console. Look on the console and see that R actually entered the commands there from your script.

If the script had one or more errors, a note from R will appear under the offending command in the console. The note will normally be helpful for fixing the script. Hopefully, the nature of the error will be evident, like a parenthesis omitted or a stray comma. Return to the R editor, fix the error(s), resave the script, and rerun it.

Let us add some things to the script. Close the window with the graph and return to the R editor window containing your script. Type in the following commands at the end of the script:

```
m=2.5*(0:100)/100
a=3.37
b=0.47
k=a*m/(b+m)
points(m,k,type="l")
```

You will recognize the statements as those adding the model equation for predator feeding rate to the graph. Save the script and run it to produce the graph of the wolf–moose data with the model equation superimposed (which appears in Chapter 1 as Figure 1.2).

Finding Errors in an R Script

R scripts for complex projects can be quite long. Seldom will such scripts run properly at first, even for R experts. Any errors in the script must be tracked down and fixed.

Tracking down errors or "bugs" in a script is called "debugging." Debugging involves some satisfying detective work.

Let us deliberately create a bug in our script to practice debugging. First, we will do some housecleaning. Make the console active and enter the following command:

```
> objects()
```

We will continue to use the convention of showing the R prompt (">") for commands issued in the console, but remember that it is not to be typed. Make sure to include the two parenthesis characters at the end. When the command is issued, you will see a list of the "objects" that your R commands have created in your workspace:

```
[1]  "a"              "b"              "k"              "kill.rate"
[5]  "m"              "moose.density"
```

These objects are stored in R and ready for future use. Now, type this command:

```
>rm(a,b,k,kill.rate,m,moose.density)
>a
```

Oops, a is no longer there, nor are any of the other objects. The command `rm()` means "remove," and it deletes from memory anything listed within its parentheses. The `objects()` and `rm()` commands are handy because wayward quantities that you create and forget about early in an R session can mess up calculations later on.

Now, we are ready to create bug havoc! Go back to the R editor. In the first command of the script, change the first letter of `moose.density` to an upper case M. Rerun the script. The graph window pops up, but no graph appears. Something has gone wrong. Look on the console to find something resembling the following statements from R (different versions of R might produce slightly different error messages):

```
> Moose.density=c(.17,.23,.23,.26,.37,.42,.66,.80,1.11,1.30,1.37,
+ 1.41,1.73,2.49)
> kill.rate=c(.37,.47,1.90,2.04,1.12,1.74,2.78,1.85,1.88,1.96,
+ 1.80,2.44,2.81,3.75)
> plot(moose.density,kill.rate,type="p")
Error in plot(moose.density, kill.rate, type = "p") :
  object 'moose.density' not found
> m=2.5*(0:100)/100
> a=3.37
> b=0.47
> k=a*m/(b+m)
> points(m,k,type="l")
```

```
Error in plot.xy(xy.coords(x, y), type = type, ...) :
  plot.new has not been called yet
```

The script has been echoed in the console along with two error messages. There are two important things to note here.

First, the plot() statement and the points() statement have the error messages, even though they were not altered, but the statement with the actual error (the upper case M) does not get flagged! In a script, an error can percolate down and cause many subsequent commands to be incorrect. Here, the command defining the Moose.density vector was a perfectly good R statement, and a peek in the console with the objects() command will establish that an object named Moose.density exists in R's memory, waiting for further duty. The problem came with the plot statement. The plot() statement was looking for a vector named moose.density, which does not exist. So, the plot() statement failed to produce a plot. Subsequently, the points() statement could not do its job, having no plot to add points to.

Second, if an error can be located and fixed, the result often is that more errors down the line become fixed. This suggests a strategy of fixing errors one by one, starting with the topmost error. It also suggests that statements or small portions of R scripts be tested as they are written, before piling further statements on (um, under!) existing statements containing errors. But even with careful and attentive construction, few R scripts of significant length run perfectly at first.

The detective work of locating elusive errors in a script should be conducted by a process of systematic search and experimentation, starting from the top. A useful tool for the task is highlighting and running portions of the script, starting with a clean workspace. Try it. First, in the console, remove all the objects in the workspace, just like you did before. Then, in the R editor, highlight the first command. In the Edit pull-down menu (and also in MS Windows in the menu resulting from right-clicking the highlighted area), find and click the option Run line or selection. The result is that the first line of the script is run. In the console, check to see which objects have been created by the command.

If the command seems to be working according to what was expected, then move down to the next statement. Highlight and run it alone and then together with its predecessors in the script.

When your bug search reaches the plot statement, the error message appears. Plot is looking for something that does not exist. It would hopefully become evident by now if it had not been discovered before this point in the search that Moose.density and moose.density are not the same objects. Users of R learn to suspect lowercase–uppercase typos and other mistypings of names as a common source of errors.

The big M is found and corrected, the script runs, and the graph is produced. Yay. But wait: Premature celebration of a completed script can be a route to disaster. Sometimes, the script will run fine, and plausible yet terribly wrong results will be produced. This happens when there is some conceptual error in the calculations themselves. R will do exactly what you tell it to do, and if what you are telling it to do is incorrect, R will faithfully carry out your instructions.

Did you enter the data correctly? Did your author type the data in this chapter faithfully from Chapter 1? To use R effectively, it helps to obsess a bit.

Sharpening Up Your Scripts with Comments

Put comments in your scripts. Any script line that begins with the sharp symbol or number symbol # is simply ignored by R. Also, if the sharp symbol occurs in the middle of a line, then the rest of the line is ignored. You can use this feature to put in explanations of the script details for the benefit of you and others who might use the script. Your author finds, for instance, that when returning to a script even a day later his memory needs to be refreshed about whatever he was trying to accomplish with each script statement.

Here is what our wolf–moose script might look like after comments are added:

```
#=======================================================
#   wolf moose graph version 20110625.R
#   R program to plot the average kill rate of moose per wolf
#   (average number of moose killed per wolf per 100 days;
#   vertical axis) with the density of moose (average number
#   per 1000 square km; horizontal axis), along with the model
#   equation for predation rate from ecological theory.
#
#   Data are from Messier, F. 1994. Ungulate population models
#   with predation: a case study with the North American moose.
#   Ecology 75:478-488.

#=======================================================

#-------------------------------------------------
#   Enter data into two vectors.
#-------------------------------------------------
moose.density=c(.17,.23,.23,.26,.37,.42,.66,.80,1.11,1.30,1.37,
1.41,1.73,2.49)
kill.rate=c(.37,.47,1.90,2.04,1.12,1.74,2.78,1.85,1.88,1.96,
1.80,2.44,2.81,3.75)

#-------------------------------------------------
#   Draw a scatterplot of the data.
#-------------------------------------------------
```

```
plot(moose.density,kill.rate,type="p")

#---------------------------------------
#  Calculate predation rate equation over a range of moose
#  densities and store in two vectors, m and k.
#---------------------------------------
m=2.5*(0:100)/100    # Range of moose densities from 0 to 2.5.
a=3.37               # Prey density at which kill is half of
                     #   maximum.
b=0.47               # Maximum kill rate.
k=a*m/(b+m)          # Model equation calculated for all values
                     #   contained in the vector m.

#---------------------------------------
#  Plot the predation rate equation data as a line plot.
#---------------------------------------
points(m,k,type="l")
```

You cannot have too many comments. Comments require more typing, but they save much mental energy in the future. Incidentally, the year–month–day method of recording dates (seen here in the first comment line as "version 20110625") gives a continually increasing number that is handy for sorting, storing, searching, and retrieving documents.

Comments are useful for debugging. You can "comment out" a line or portion of a script to omit that portion when the script is run. Just insert the sharp sign at the beginning of each line that is to be omitted from running. You can comment out a portion of a script and then substitute some alternative statement or set of statements for running as a debugging experiment, without losing all your original typing.

Real-World Example

Let us return to finances for awhile. Suppose you are ready to buy your first house, a mobile home, for $30,000. You know you qualify for a small mortgage, but you would like to know how big the monthly payments will be.

You will provide a down payment of $6000. There will be about $1000 of extra fees, which will be added to the amount that you borrow, and so the amount of money borrowed for the mortgage will total $25,000. This is the "principal" of the loan. The mortgage provider agrees to a 5% annual interest rate on any unpaid balance of the principal. The problem is to determine a monthly series of payments that gradually pay off the balance and all interest by the end of 30 years (or 360 months, a standard mortgage length, at a monthly interest of $5\% \div 12 = 0.417\%$).

There is a somewhat complicated-looking equation for calculating the monthly payment. Everyone who goes to business school learns the equation. Here is the equation written for our loan details:

$$\text{Payment} = \frac{(1+0.00417)^{360}(0.00417)(25{,}000)}{(1+0.00417)^{360}-1}.$$

If you are the sort of person that gets frustrated from not understanding how such an equation arises, the accompanying box (Deriving the Monthly Loan Payment Equation) provides all the algebraic steps of the derivation. The derivation is long but not difficult. To go through the derivation, take a sheet of paper and write out and understand every step. The effort will be worth it (and if you know this material, you will breeze through business school). Just skip the box if you want to go straight to the R calculations.

Go to the R console and calculate that monthly payment! Our R command, and the result, would look something like this:

```
> (1+.00417)^360*.00417*25000/((1+.00417)^360-1)
[1] 134.2665
```

The lender will want $134.27 every month for 360 months.

Now, the payment calculation for our particular mortgage is hardly worth an R script because it takes only one line in the console. What would be useful, however, is a script to calculate all the payment details, for a loan of any principal amount, for a term of any number of months, or for an annual interest rate of any amount. If you had that at your fingertips, you would be almost ready to open your own bank.

In the "Deriving the Monthly Loan Payment Equation" box, the derivation provided us with two other equations besides that for the monthly payment: the amount of principal paid in each month and the amount of principal remaining unpaid. The symbols defined in the box were as follows: m = total number of months (payments) in the loan, s = monthly interest rate (annual rate ÷ 12), and P = principal amount borrowed. In terms of those symbols, the three loan equations are as follows:

$$\text{Monthly payment} = \frac{(1+s)^m \, sP}{(1+s)^m - 1},$$

$$\text{Principal paid in month } t = \frac{(1+s)^{t-1} \, sP}{(1+s)^m - 1},$$

$$\text{Principal remaining after month } t = P\left[1 - \frac{(1+s)^t - 1}{(1+s)^m - 1}\right].$$

Let us plan an R script. What do we want it to do? Calculating the monthly payment amount would be useful, of course. Also, calculating the principal paid in every month seems useful. Note that the principal paid is a quantity that changes each month—this seems highly suited for the vector capabilities of R, if we are clever enough to formulate the script that way (really, we do not want to write 360 lines to calculate each month separately!). Knowledge of the amounts of the principal payments would then give us the interest amounts for each month. And, how about producing a graph of, say, the amount of unpaid principal remaining in the loan every month? As well, let us add up all the interest paid throughout the life of the loan, which is likely to be a shocker.

It is helpful to write out the sequential scheme for all the tasks before starting any script commands.

Step 0. Assign numerical values to P (principal), m (total number of monthly payments), and i (annual interest rate). Calculate s (monthly interest rate). These will be placed at the front of the script so that a user of the script can easily find and change them.

Step 1. Calculate a vector of values of time (months) going from 1 to m.

Step 2. Calculate the monthly payment (a single number) using the first loan equation.

Step 3. Calculate a vector of principal amounts of the loan paid each month using the second loan equation.

Step 4. Calculate a vector of principal amounts of the loan remaining unpaid each month using the third loan equation.

Step 5. Calculate a vector of the interest amounts paid each month by subtracting the principal amounts paid from the monthly payment.

Step 6. Calculate the total interest paid by summing all the interest amounts paid each month. There is a resource in R that can make this calculation easy: if x is a vector, then sum(x) adds all the elements in x.

Step 7. Print the results to the console.

Step 8. Draw a graph of the principal remaining in the loan each month (vertical axis) versus the vector of months (horizontal axis).

Ambitious? Sure! R is all about thinking big. My advice now is for you to open a clean, fresh R editor for a new script and try your hand at writing R statements for each step above, one at a time. After every step, compare your statements to the corresponding statements written below. Remember that there might be several different correct ways of writing an R statement to do a calculation. Your way might be as good or better than the corresponding one below.

After I chart out the steps for a script, I like to use the steps as comments! If you just want to see calculated results quickly, skip the typing of

the comments, but be sure to put some comments back in the version of the
script you save for long-term use. Here we go:

```
#================================================
# loan payment.R:  R script to calculate and plot monthly loan
# payment information.

#================================================

#------------------------------------------------
# Step 0.  Assign numerical values to P (principal), m (total
# number of monthly payments), and i (annual interest rate).
# Calculate s (monthly interest rate).
#------------------------------------------------
P=25000
m=360
i=.05     # Interest is 100*i percent per year.
s=i/12    # Monthly interest rate.

#------------------------------------------------
# Step 1.  Calculate a vector of values of time (months) going
# from 1 to m.
#------------------------------------------------
t=1:m

#------------------------------------------------
# Step 2.  Calculate the monthly payment (a single number)
# using the first loan equation.
#------------------------------------------------
monthly.payment=(1+s)^m*s*P/((1+s)^m-1)

#------------------------------------------------
# Step 3.  Calculate a vector of principal amounts paid each
# month of the loan using the second loan equation.
#------------------------------------------------
principal.paid.month.t=(1+s)^(t-1)*s*P/((1+s)^m-1)

#------------------------------------------------
# Step 4.  Calculate a vector of principal amounts remaining
# unpaid each month of the loan using the third loan equation.
#------------------------------------------------
principal.remaining=P*(1-((1+s)^t-1)/((1+s)^m-1))

#------------------------------------------------
# Step 5.  Calculate a vector of the interest amounts paid
# each month by subtracting the principal amounts paid from the
# monthly payment.
#------------------------------------------------
interest.paid.month.t=monthly.payment-principal.paid.month.t

#------------------------------------------------
# Step 6.  Calculate the total interest paid by summing all the
# interest amounts paid each month using the sum( ) function in R.
#------------------------------------------------
```

```
total.interest.paid=sum(interest.paid.month.t)

#------------------------------------------------
# Step 7.  Print the results to the console.
#------------------------------------------------
monthly.payment
total.interest.paid
t
principal.paid.month.t
interest.paid.month.t
principal.remaining

#------------------------------------------------
# Step 8.  Draw a graph of the principal remaining in the loan
# each month (vertical axis) versus the vector of months
# (horizontal axis).
#------------------------------------------------
plot(t,principal.remaining,type="l")
```

Save your script, run it, debug it, run it again, and debug it again. Compare the R commands with the corresponding loan equations closely and make sure you are comfortable with the rules of the R syntax (review the rules in Chapter 1 if necessary). When you have it right, the script will print voluminous numbers to the console and will produce the graph in Figure 2.1.

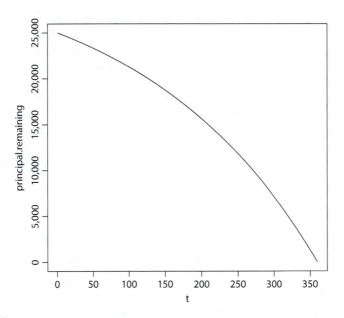

FIGURE 2.1
Principal remaining (vertical axis) at month *t* (horizontal axis) in a loan of $25,000 for 360 months (30 years) with an annual interest rate of 5%.

You might have noticed that the monthly loan payment calculated by the script is \$134.2054, which differs slightly from the figure of \$134.2665 that we calculated with one line at the console. That earlier calculation used a rounded monthly interest rate of $0.05/12 \approx 0.00417$ instead of the double precision approximation to $0.05/12$ that is produced and used in the script. If you just calculate $0.05/12$ at the console, R will round the figure for printing to 0.004166667. Remember, the "floating point arithmetic" used by all calculators and computers results in round-off error. The errors will be propagated throughout as lengthy calculations, and it is usually best to avoid rounding at the beginning, rounding if needed only at the end.

In addition to the monthly payment, at the beginning of the numbers printed to the console was the total interest paid over the duration of the loan. It is daunting to realize that the total interest paid over the duration of the loan will almost equal the principal. You can try rerunning your script now to see what effect a shorter loan period, like 15 years instead of 30, will have on the payment schedule.

Final Remarks

Because `t`, `principal.paid.month.t`, `interest.paid.month.t`, and `principal.remaining` in the above example are vectors each with 360 (or m) elements, printing them to the console produced an avalanche of numbers. It would be better if we could organize these into a table of some sort for nicer printing. As well, we will want to have ways to bring large tables of data into R for analysis. Chapter 5 discusses data input and output.

You might be intrigued by the advent, in the script given earlier, of the function called `sum()` that adds up the elements of any vector. R has many such functions, and Chapter 3 will introduce a number of them. Also, you will learn in Chapter 3 how to write your own functions. Or perhaps more importantly, you will learn why you would want to write your own functions!

You might have noticed by now that an R script is remarkably similar to what computer scientists would call a computer program and that writing a script suspiciously resembles computer programming. Yes, that is right. You: geek! Get used to it.

DERIVING THE MONTHLY LOAN PAYMENT EQUATION

Here, we will figure out what the loan payment equation is, starting from scratch! Getting there might seem like it consumes lots of algebra, but be assured that each step is small (and that algebra is a renewable resource). You can skip this box and just accept the loan payment equation if you want to continue with the R example. If you decide to continue thinking inside the box, pause here to buy stock in a pencil company.

Before we launch into the derivation of the loan payment equation, we need to know about a cute math result called the geometric series formula that will help us obtain the equation. Take any real number that is not zero, say r, and add up all its powers, up to, say, r^k, where k is any positive integer:

$$1 + r^1 + r^2 + r^3 + \cdots + r^{k-1} + r^k.$$

The 1 in the sum is, of course, r^0. The sum is called a geometric series. Now, suppose we multiply the whole sum by the quantity $(1-r)/(1-r)$. That is, just multiply the sum by 1, so the value of the sum will not change. The result will be a fraction:

$$\frac{(1-r)\left(1 + r^1 + r^2 + r^3 + \cdots + r^{k-1} + r^k\right)}{(1-r)}.$$

Look at what the product in the numerator will become: each term of the geometric series will be multiplied by 1 and also by $-r$:

$$(1-r)\left(1 + r^1 + r^2 + r^3 + \cdots + r^{k-1} + r^k\right)$$
$$= (1)\left(1 + r^1 + r^2 + r^3 + \cdots + r^{k-1} + r^k\right) + (-r)\left(1 + r^1 + r^2 + r^3 + \cdots + r^{k-1} + r^k\right)$$
$$= \left(1 + r^1 + r^2 + r^3 + \cdots + r^{k-1} + r^k\right) + \left(-r^1 - r^2 - r^3 - \cdots - r^k - r^{k+1}\right).$$

All the terms cancel except the 1 and the $-r^{k+1}$, leaving just $\left(1 - r^{k+1}\right)$ in the numerator.

Remembering the $1-r$ term in the denominator, we have derived a remarkable simplifying result for a sum of powers of r:

$$1 + r^1 + r^2 + r^3 + \cdots + r^{k-1} + r^k = \frac{1 - r^{k+1}}{1-r}.$$

This formula is the geometric series formula, and it appears in all sorts of financial and scientific calculations. You should commit it to memory (your brain's hard drive, not its RAM).

We return to our mortgage problem. We should pose the problem of monthly payments using months as the timescale. So, there are $30 \times 12 = 360$ total payments in the mortgage. The annual interest rate should be expressed as a fraction for calculations, so we take the annual interest rate to be 0.05. Then, the monthly interest rate is $0.05/12 = 0.00417$ (after rounding a bit).

Each month, you will pay some of the principal, plus a month of interest on the remaining principal. The amount of principal paid each month will be different. Let us write $x_1, x_2, x_3, \ldots, x_{359}, x_{360}$ for the monthly principal amounts paid. They are all unknown at this point! But what we do know is that they all must add up to the principal amount of the loan:

$$x_1 + x_2 + x_3 + \cdots + x_{359} + x_{360} = 25{,}000.$$

Also, we know something about what each monthly payment will look like. The first month's payment will be the principal amount x_1 plus the interest amount given by $(0.00417)(25{,}000)$. The second payment will be the principal amount x_2 plus the interest amount given by $(0.00417)(25{,}000 - x_1)$. The third monthly payment will be x_3 plus $(0.00417)(25{,}000 - x_1 - x_2)$. In each month's loan payment, the unpaid principal gets reduced by the sum of the principal payments up to that month:

Month 1: $x_1 + (0.00417)(25{,}000),$

Month 2: $x_2 + (0.00417)(25{,}000 - x_1),$

Month 3: $x_3 + (0.00417)(25{,}000 - x_1 - x_2),$

$\qquad \vdots \qquad\qquad \vdots$

Month 360 : $x_{360} + (0.00417)(25{,}000 - x_1 - x_2 - x_3 - \cdots - x_{359}).$

As well, we want each month's total payment to be the same. Look at the stack of payments above and subtract the first payment from the second payment. If the payments are the same, the result of the subtraction will be zero. But the result also reveals how x_2 is related to x_1:

$$x_2 - x_1 - (0.00417)x_1 = 0,$$

or, putting the terms with x_1 on the right-hand side,

$$x_2 = (1 + 0.00417)x_1.$$

Do the same with the second and third payments. Subtract the second from the third to get

$$x_3 = (1 + 0.00417)x_2,$$

or, substituting for x_2,

$$x_3 = (1+0.00417)(1+0.00417)x_1 = (1+0.00417)^2 x_1.$$

Does the pattern look familiar? The principal payments grow geometrically, just like your money in the bank, as we calculated in Chapter 1. In general, the principal payment in month t will be related to the first principal payment by

$$x_t = (1+0.00417)^{t-1} x_1.$$

All we need to know now is how much the first principal payment must be and everything else can be calculated from that. But remember, the principal payments must all add up to the total principal (total amount borrowed):

$$x_1 + x_2 + x_3 + \cdots + x_{359} + x_{360} = 25{,}000.$$

Substitute for x_2, x_3, ..., x_{359}, x_{360} to get

$$x_1 + (1+0.00417)x_1 + (1+0.00417)^2 x_1 + \cdots + (1+0.00417)^{359} x_1 = 25{,}000.$$

We are almost there. Factor out the x_1, use our geometric series formula for the sum of powers, and solve for x_1:

$$x_1[1+(1+0.00417)^1 +(1+0.00417)^2 + \cdots +(1+0.00417)^{359}] = 25{,}000,$$

$$x_1 \frac{1-(1+0.00417)^{360}}{1-(1+0.00417)} = 25{,}000$$

(in the denominator, $1-(1+0.00417) = -0.00417$),

$$x_1 \frac{1-(1+0.00417)^{360}}{-0.00417} = 25{,}000$$

(multiply numerator and denominator by –1),

$$x_1 \frac{(1+0.00417)^{360} -1}{0.00417} = 25{,}000 \text{ (solve for } x_1 \text{)},$$

$$x_1 = \frac{(0.00417)(25{,}000)}{(1+0.00417)^{360} -1}.$$

Thus, the principal payment in month t is

$$x_t = (1+0.00417)^{t-1} \frac{(0.00417)(25{,}000)}{(1+0.00417)^{360}-1}.$$

The unpaid principal remaining in the loan after the payment at month t is

$$25{,}000 - (x_1 + x_2 + x_3 + \quad + x_t),$$

$$= 25{,}000 - x_1[1+(1+0.00417)^1+(1+0.00417)^2+ \quad +(1+0.00417)^{t-1}]$$

$$\text{(use the geometric series formula)},$$

$$= 25{,}000 - x_1 \frac{1-(1+0.00417)^t}{1-(1+0.00417)} \quad \text{(substitute for } x_1),$$

$$= 25{,}000 - \frac{(0.00417)(25{,}000)}{(1+0.00417)^{360}-1}\left(\frac{1-(1+0.00417)^t}{1-(1+0.00417)}\right)$$

$$\text{(factor out 25,000 and note that } 1-(1+0.00417) = -0.00417),$$

$$= 25{,}000\left[1 - \frac{1-(1+0.00417)^t}{1-(1+0.00417)^{360}}\right].$$

And finally, the total monthly loan payment (principal plus interest) can simply be found from the formula for the first month because all the total monthly payments are the same:

$$\text{Monthly payment} = x_1 + (0.00417)(26{,}000)$$

$$= \frac{(0.00417)(26{,}000)}{(1+0.00417)^{360}-1} + (0.00417)(26{,}000).$$

The above is the monthly payment equation. It has many algebraic variants. We can rearrange it slightly to get it in a form often seen in business texts, by combining the two terms using a common denominator:

$$\text{Monthly payment} = \frac{(0.00417)(26{,}000)}{(1+0.00417)^{360}-1}$$

$$+ \frac{\left[(1+0.00417)^{360}-1\right](0.00417)(26{,}000)}{(1+0.00417)^{360}-1}$$

$$= \frac{(1+0.00417)^{360}(0.00417)(26{,}000)}{(1+0.00417)^{360}-1}.$$

In the algebra above, we were careful to carry all the individual numbers through to the very end. If we repeated all the algebra using symbols, say, m = total number of months (payments) in the loan, s = monthly interest rate (annual rate ÷ 12), and P = principal amount borrowed, we would arrive at

$$\text{Monthly payment} = \frac{(1+s)^m \, sP}{(1+s)^m - 1},$$

$$\text{Principal payment in month } t = x_t = \frac{(1+s)^{t-1} \, sP}{(1+s)^m - 1},$$

$$\text{Principal remaining after month } t = P\left[1 - \frac{(1+s)^t - 1}{(1+s)^m - 1}\right].$$

Try the algebra from the beginning using symbols!

WHAT WE LEARNED

1. An R script is a prepared list of R commands that are processed sequentially by R from top to bottom. In the Windows and Mac versions of R, a script can be prepared, saved, and run using a text editor provided in R called the R editor. In the current Unix version of R, one uses any outside text editor to prepare and save the script.

2. The R editor is accessed from the `File` menu: `File→New script` to open a blank editor for a new script, `File→Open script` to open an existing script.

3. An R script in the R editor is saved as a file from the `File` menu: `File→Save` for changes to an already existing file, `File→Save as...` for a first-time save. The usual extension for names of files containing R scripts is ".R." The R editor does not automatically add the extension when a file is newly saved.

4. "Running" an R script means sending the R commands to the console for execution. In the R editor, a script can be run using the `Edit` menu: `Edit→Run all`. In Unix, a script is run by copying the entire R script and pasting it into the R console at the active prompt, or by using the `source(" ")` command in the console with the folder directory and file name entered between the quotes.

5. Portions of an R script can be run by highlighting them in the R editor and then using the `Edit` menu: `Edit→Run line or`

selection. In Unix, copy and paste the desired portion into the R console at the active prompt.

6. An R script will often have errors or "bugs" at first due to mistypings and/or errors in formulating the calculation problem for R. Running a script line by line from the beginning is a useful way of tracking down bugs. Bugs can cause subsequent R statements in the script to be incorrect, and fixing one bug often fixes several later bugs.

7. Comments are lines in a script that are ignored by R. Comments serve as valuable explanations and memory aids, and liberal use of comments is recommended. Comments are preceded by a sharp symbol or number symbol #. Anything to the right of the # symbol on a line is ignored as part of the comment.

Example:

```
# Here is an illustration of element-by-element vector
#   multiplication in R.
x=c(1,2,3)      # x is defined as the vector with elements
                # 1, 2, 3.
y=c(-1,-2,-3)   # y is defined as the vector with elements
                # -1, -2, -3.
z=x*y           # z is defined as the vector containing the
                # elements of x multiplied by the
                # corresponding elements of y.
z               # Print z to the console.
```

Computational Challenges

The following are some of the challenges from Chapter 1. Here, your challenges are to set up these problems as R scripts for calculation in R. I hope you will agree that the R commands for accomplishing Chapter 1 challenges might best be saved for future use and reference in the form of R scripts. Did you save your commands somehow after doing the Chapter 1 challenges? That might have come in handy now!

2.1. Evaluate the following expressions. Do them all with one script.

$$\frac{93^2 - 164}{46^3 + 189} \qquad 376 - \frac{23^2}{4} \qquad \frac{59 + 48^2}{-9 + 22^2} - \frac{-16 + 55^2}{13 + 29^2} \qquad 18^4 - 16^3 + 14^2 - 12$$

$$3^x \quad \text{for} \quad x = 1, 2, ..., 20 \qquad 5^x \quad \text{for} \quad x = 1, 2, ..., 10$$

2.2. Draw the investment equation (money in the CD) again, except use a much longer time horizon. Allow time to range as many years into the future (50?) as, say, you have until you reach the retirement age of 65. Gaze at the graph with awe. Yes, it is really your choice: having those fancy designer jeans now or having many times the sale price later!

Add two or three more curves onto your graph showing the effects of two or three different interest rates.

2.3. The following are the population sizes of the United States through its entire history, according to the U.S. Census. Construct a line plot (type="l") of the U.S. population (vertical axis) versus time (horizontal axis). By the way, rounding the population sizes to the nearest 100,000 or so will hardly affect the appearance of the graph.

1790	3,929,214
1800	5,308,483
1810	7,239,881
1820	9,638,453
1830	12,860,702
1840	17,063,353
1850	23,191,876
1860	31,443,321
1870	38,558,371
1880	50,189,209
1890	62,979,766
1900	76,212,168
1910	92,228,496
1920	106,021,537
1930	123,202,624
1940	142,164,569
1950	161,325,798
1960	189,323,175
1970	213,302,031
1980	236,542,199
1990	258,709,873
2000	291,421,906

Change the script and repeat the plot six more times (saving the graph each time), using type="p", type="b", type="c", type="o", type="h", and type="l". Compare the different graph types. What different aspects of the data are emphasized by the different graph types?

2.4. If you throw a baseball at an angle of 45°, at an initial velocity of 75 mph, while standing on a level field, the ball's horizontal distance x traveled after t seconds is described (neglecting air resistance) by the following equation from Newtonian physics:

$$x = 27.12t.$$

Furthermore, the height above the ground after t seconds, assuming the ball was initially released at a height of 5 ft, is described by

$$y = 1.524 + 19.71t - 4.905t^2.$$

The equations have been calibrated to give the distance x and height y in meters. The ball will hit the ground after about 4.09 seconds. Calculate a vector (say, x) of baseball distances for a range of values of t from 0 to 4.09. Calculate a vector of baseball heights (say, y) for the same collection of times. Make a plot of x (horizontal axis) and y (vertical axis). Read from the graph of the ball's trajectory how high and how far, approximately, the ball will travel.

NOTE: For different initial throwing velocities and angles, the above baseball equations will have different numerical coefficients in them. The equations can be written in more general form to accept different initial conditions, but to do that we will need a small bit of trigonometry (Chapter 9).

2.5. According to Newton's universal law of gravitation, the acceleration of an object in the direction of the sun due to the sun's gravity can be written in the form

$$a = \frac{1}{r^2}.$$

Here, r is the distance of the object from the sun's center, in astronomical units (AU) of distance. One AU is the average distance of the Earth from the sun, about 150 million kilometers. The units of a are scaled for convenience in this version of Newton's equation so that one unit of acceleration is experienced at a distance of 1 AU. Use the equation to calculate the gravitational accelerations at each of the planets' average distances from the sun:

Planet:	Mercury	Venus	Mars	Jupiter	Saturn	Uranus	Neptune	(Pluto)
Distance:	0.39	0.723	1.524	5.203	9.539	19.18	30.06	39.53

Pluto is now considered to be a large comet-like object or dwarf planet that originated from the Kuiper Belt.

2.6. Using the equation from Question 2.5, draw a graph of the gravitational acceleration (vertical axis) versus a range of values of r ranging from around 0.4 AU (distance of Mercury) to around 5.2 AU (distance of Jupiter). According to Newton's gravitational law, is there any distance at which the sun's gravity is escaped entirely?

2.7. Using the kill rate equation for the wolf–moose system that you studied in Chapter 1, do some numerical exploration/experimentation to find out where the curve levels off (the maximum kill rate of the average wolf). How is it possible that some of the real data had kill rate values above that maximum? Does the leveling off point resemble any numerical quantity you see in the equation itself?

2.8. To decrease the use of insecticides in agriculture, predator insects are often released to combat insect pests. Coccinellids (lady beetles) in particular have a voracious appetite for aphids. In a recent study (Pervez and Omkar 2005), entomologists looked at the suitability of using coccinellids to control a particular aphid, *Myzus persicae* (common name is the "green peach aphid"), a serious pest of many fruit and vegetable crops. In the study, the entomologists experimentally ascertained aphid kill rates for three different species of coccinellids:

APHID DENSITY (#/jar)	COCCINELLID FEEDING RATE (# eaten per 24 hr)		
	Cheilomenes sexmaculata	*Coccinella transversalis*	*Propylea dissecta*
25	21	21	15
50	37	37	26
100	65	60	42
200	102	90	59
300	125	109	69
400	141	120	74
500	154	129	79
600	164	136	82
700	170	140	83
800	177	143	85

Enter the data columns above into vectors, giving them descriptive names. For each type of coccinellid, use R to construct a scatterplot (type="p") of the feeding rate of the coccinellid versus aphid density. Then, add a kill rate curve to the coccinelid/aphid graph. Use the following constants in the kill rate equations:

$$C.\ sexmaculata:\ a = 234.5,\ b = 261.9,$$
$$C.\ transversalis:\ a = 178.9,\ b = 194.9,$$
$$P.\ dissecta:\ a = 100.4,\ b = 139.8.$$

Save and close each graph before starting the graph for the next coccinellid.

Reference

Pervez, A., and Omkar, A. 2005. Functional responses of coccinellid predators: An illustration of a logistic approach. *Journal of Insect Science* 5:1–6.

3

Functions

Sometimes there are calculation tasks that must be performed again and again. R contains many useful calculation tasks preprogrammed into functions. You can use these functions in the command console or in any scripts.

A function usually is designed to take a quantity or a vector, called the argument of the function, and calculate something with it. The function returns the value of the calculation as output for further use.

For instance, a simple function is the sum() function. Type the following R commands in the console:

```
> x=c(5.3,-2.6,1.1,7.9,-4.0)
> y=sum(x)
> y
[1]  7.7
```

In the above commands, the vector x is the **argument** of the function and y is the value returned by the function. The sum() function adds up the elements of any vector used as its argument.

Here are some functions in R that can come in handy:

length(x) returns the number of elements in the vector x

sum(x) returns the sum of the elements in x

prod(x) returns the product of the elements in x

cumsum(x) returns the cumulative sum of the elements in x

cumprod(x) returns the cumulative product of the elements in x

mean(x) returns the arithmetic average of the elements in x (equivalent to sum(x)/length(x))

abs(x) returns a vector containing the absolute values of the elements in x

sqrt(x) returns a vector containing the square roots of the elements in x (identical to x^.5)

Fire up R and try them out:

```
> x=c(2,3,5,1,0,-4)
> length(x)
[1] 6
> sum(x)
```

```
[1]  7
> prod(x)
[1]  0
> cumsum(x)
[1]   2   5 10 11 11   7
> cumprod(x)
[1]   2   6 30 30   0   0
> mean(x)
[1] 1.166667
> abs(x)
[1] 2 3 5 1 0 4
> sqrt(x)
[1] 1.414214 1.732051 2.236068 1.000000 0.000000      NaN
Warning message:
In sqrt(x) : NaNs produced
```

Whoops, something went wrong in the square root function. An inspection of the vector x reveals that the last element has the value −4. The square root of a negative number does not exist as a real number. R printed the character "NaN" for the result, which stands for "not a number." Subsequent calculations in which one or more of the input numbers is a NaN result in a NaN as the output. For instance, try,

```
> y=c(-2,4,7,9,11,3)
> sqrt(x)*sqrt(y)
[1]      NaN 3.464102 5.916080 3.000000 0.000000      NaN
Warning messages:
1: In sqrt(x) : NaNs produced
2: In sqrt(y) : NaNs produced
```

In y, the first element is negative and does not have a square root, just like the last element in x. The multiplying of the corresponding elements of sqrt(x) and sqrt(y) produced NaNs whenever either element was a NaN.

In R, you can use functions freely in assignment statements. You can even use functions or R calculations as arguments in other functions. For instance, Pythagoras' theorem states that the length of the hypotenuse of a right triangle is equal to the square root of the sum of the squares of the other two sides. Suppose the two lesser sides of a right triangle have lengths 12.3 and 20.5, a piece of cake in R. First, visualize the calculation in an equation for the hypotenuse length h:

$$h = \sqrt{(12.3)^2 + (20.5)^2}.$$

If we put the two side lengths into a vector named, say, sides, then the R commands might look like the following:

```
> sides=c(12.3,20.5)
> h=sqrt(sum(sides^2))
> h
[1] 23.90690
```

Here, `sides` is the vector given by (12.3, 20.5), `sides^2` is the vector [(12.3)², (20.5)²], `sum(sides^2)` is then the quantity $(12.3)^2 + (20.5)^2$, and `sqrt(sum(sides^2))` takes the square root of the result.

A function in R might also produce something nonnumerical from its argument, like a table or a graph. In this respect, a function in R differs from the usual definition of a mathematical function, which takes as its argument a numerical value (or a set of numerical values) and from the argument calculates only a numerical value (or a set of numerical values). I will sometimes use the term "R function" to distinguish such an entity in R from a mathematical function. We have seen, for instance, the `plot()` function that produces an *x-y* graph.

In later chapters, we will study many additional special functions that are built in R.

Creating New Functions in R

R allows you to build your own functions. This especially handy feature can be a great time saver for calculation tasks in, say, science homework, which use the same formula but are done many times with different numbers as input.

Here is how you build your own R function. We will borrow Pythagoras' hypotenuse calculation discussed earlier as the example. Open the R editor window and enter the following script:

```
length.hyp=function(x) {
  h=sqrt(sum(x^2))
  return(h)
}
```

Now run the script (Remember? Click Edit → Run all). You will see that nothing much happens. But now, go to the console and enter the following commands:

```
> sides=c(3,4)
> length.hyp(sides)
[1] 5
```

If the two shorter sides have lengths 3 and 4, then the hypotenuse has length 5. There is a lot going on here; let us review all these R statements.

The first statement in the script above defines `length.hyp()` as an R function that will take the argument *x*. The word `length.hyp` is just a name that we chose. We could have named the function `lh()`, say, for less typing, but descriptive names are easier to follow when writing and debugging lengthy scripts. Also, choosing names that already exist in R, such as `length`, should generally be avoided.

One or more R statements that define the actual calculations then appear between the curly braces "{ }". The defining R statements take whatever numbers are in x and calculate whatever quantities or items we desire as the output of the function. In our function, we calculate a quantity h, seen to be the hypotenuse length resulting from any two side lengths contained in x. The `return(h)` statement then defines the output of the function to be whatever number or numbers are in h. If we had defined h to be a vector, then the output of the function would be a vector.

In the commands that you subsequently entered in the console, `sides` was defined as a vector containing the numbers 3 and 4. Then, you invoked the newly created R function called `length.hyp()` to do the hypotenuse length calculation using the vector `sides`. R printed the familiar result from a 3,4,5 triangle.

There are three important things to note about this function definition:

1. In the script defining the function `length.hyp()`, the symbol x is a placeholder. It did not have to be predefined to contain numbers; the script ran just fine. Instead, the function "lives" in the workspace and will take any vector as an argument and attempt to do to it whatever the function's defining statements do to x. By the way, we could have used a different name than x for the argument if we wished.

2. The values of x and h inside the function are "local" to the function. These local vectors exist in a special place in the computer memory set aside just for the function, and their values are not available outside the function. If you type "h" at the console, you will see that no entity called h exists in your workspace.

3. Although the function `length.hyp()` now exists in the workspace, just like `sum()` or `sqrt()`, the new function is not permanent. It will disappear when R is turned off, unless the workspace is saved and then reloaded the next time you use R. You can have many saved workspaces, perhaps one for chemistry, one for math, and so on. However, remembering the contents of different workspaces can become confusing. An alternative is simply to save your function scripts under descriptive file names and then run whichever ones you need each time you use R.

More about User-Defined R Functions

Note how the curly braces and defining statements were positioned typographically in the function definition. The function did not have to be typed that way. You could have typed it as follows:

```
length.hyp=function(x)
{
h=sqrt(sum(x^2))
return(h)
}
```

Although these statements would work just fine, the former way uses a typographical convention that many programmers prefer. Everything between the function name and the closing curly brace is indented a bit. This indentation highlights all the statements within the function. In large R scripts for scientific projects, one might define many functions. Debugging such a script could require chasing the values of many local and global vectors through an intricate series of calculations. Seeing at a glance which R statements are included inside a function can help such a chase tremendously.

The ordinary vectors and quantities defined outside of R functions are called "global" quantities. For instance, sides above was a global vector. Global quantities that exist in the workspace can be used or referenced inside the statements that define R functions. However, using global quantities inside functions is not considered a good programming style. During an R session, the values of global quantities might be changed in the course of the calculations, and a function that depended on a previous global value might cause errors. An improved, more transparent practice is to bring global quantities into the function only as arguments to the function. The practice allows the effect of global quantities on the output of the function to be more easily traced when debugging.

Of course, instead of writing an R function, you could just write a script to do the desired calculations, with assignment statements at the beginning where you would put the input numbers. Our mortgage script in Chapter 2 was an example. Just change the number of months, the interest rate, and so on and rerun the script to get new results.

R functions, however, provide more convenience. First, after you run a function, it is "alive" in your workspace, ready to be used. Run it at the beginning of your homework session, and you can use the function repeatedly in your console, putting in the new numbers for every new homework problem. No need to open/change/save/run a script each time.

Second, the R functions you build can be used inside other R functions and/or in R scripts. Complex calculations can be conveniently organized.

Third, your workspace grows in the sophistication of its capabilities as you add R functions. You can write and share functions with colleagues. You can use R functions from students in previous courses and create R functions for future students. You can post them and improve them.

R grows on itself, if you get in the right spirit. That spirit includes writing your own functions to take care of complex calculations that you need to do often. That spirit includes writing functions to make life easier for future students and future scientists!

Real-World Example

Generations of students have struggled through various calculations in basic chemistry. Even on a calculator, the calculations can be laborious. I learned these calculations for the first time on a slide rule, and a particularly large problem set assigned by a particularly stern teacher convinced me to avoid chemistry for many years.

Let us release future chemistry students from numerical drudgery and write an R function to perform a frequent chemistry calculation. We will build a function to calculate the mass of a standard amount of any chemical compound.

A **mole** of an element or compound is defined to contain 6.02×10^{23} particles of that element or compound (atoms in the case of an element; molecules in the case of a compound). That rather large number is known as Avogadro's number. It was picked by chemists as a sort of standard amount of a substance and represents the number of atoms in 12 g of pure carbon-12 (think of a little pile of charcoal powder weighing a little less than five U.S. pennies). The **atomic mass** of an atom is the mass in grams of a mole of those particular atoms. A carbon-12 atom has an atomic mass of 12, and so a mole of carbon-12 atoms would weigh 12 g at sea level on Earth. One atom of hydrogen-1 has an atomic mass of 1, and so a mole of pure hydrogen-1 (with no isotopes such as deuterium or tritium) would weigh 1 g at sea level.

In the Earth's crust and atmosphere, some heavier or lighter isotopes of most elements occur naturally in small amounts, and so on average, a mole of everyday unpurified carbon atoms would have a mass of around 12.01 g, due to trace amounts of heavier carbon (especially carbon-14). A mole of everyday hydrogen atoms would have a mass of 1.008 g. The numbers 12.01 and 1.008 (without measurement units) are called the **atomic weights** of carbon and hydrogen respectively. Expressed as grams per mole, the quantities 12.01 and 1.008 are called the **molar masses** of carbon and hydrogen, respectively.

In general, the molar mass of an element or compound gives the average mass in grams of 1 mol of that substance, taking into account the average abundance of isotopes under normal conditions found on Earth.

The masses of other elements and compounds, of course, are different from that of carbon. To calculate the mass of a mole of, say, water, we must find the atomic weight of one molecule of water. A water molecule has two hydrogen atoms and one oxygen atom, and we need to combine their atomic weights. Atomic weights are commonly listed in chemistry tables (online and in textbooks). From such a table, we find that hydrogen has an atomic weight of 1.008 and oxygen has an atomic weight of 16.00. Can you see that the mass of 1 mol of water would be obtained by the following calculation?

$$2(1.008) + 1(16.00).$$

The calculation would be a simple one in the R console. Try it.

An R function would make the calculation easier to do. Furthermore, if we are clever, we can build the function to calculate the molar mass of any compound, not just water.

The key to calculations in R, you will realize if you are catching on, is to find vectors in the calculations. For water, notice that we had two atoms of hydrogen and one atom of oxygen. Try thinking of those numbers as a vector. At the console, type

```
> num.atoms=c(2,1)
```

Notice also that the corresponding atomic weights were 1.008 and 16.00. Let us try thinking of the weights as a vector too. Type

```
> atomic.weights=c(1.008,16.00)
```

If you are getting into the R way of thinking, you will notice that the calculation in the square brackets of the atomic weight expression above can be written in R as

```
> molar.mass=sum(num.atoms*atomic.weights)
> molar.mass
[1] 18.016
```

A mole of water weighs about 18 g at sea level.

Inside the sum() function above, each element of num.atoms is multiplied by the corresponding element of atomic.weights (take note that the word "element" has different meanings in R and chemistry). That gives us the key R insight for building our function: the form of the calculation would be the same for any compound, provided that the numbers of atoms of each element were in num.atoms and corresponding weights were in atomic.weights. Let us try this idea as a function. We will redefine molar.mass as a function that takes two arguments, num.atoms and atomic.weights.

First, do some housekeeping at the console and clean the workspace a bit:

```
> rm(molar.mass)
```

Next, enter the following function definition in the R editor and run it:

```
molar.mass=function(x,y) {
  mm=sum(x*y)
  return(mm)
}
```

Here, molar.mass is now redefined as an R function. Notice that the function has two arguments, x and y. Each of these is a vector, and each will now be local within the function. The first vector x will contain the number of atoms of each element in a molecule of the compound, and the second vector y will contain the corresponding atomic weights of those elements.

Go to the R console now and try out a few compounds. Try water:

```
> num.atoms=c(2,1)
> atomic.weights=c(1.008,16.00)
> molar.mass(num.atoms,atomic.weights)
[1] 18.016
```

Try sucrose (table sugar: 12 carbons, 22 hydrogens, 11 oxygens):

```
> num.atoms=c(12,22,11)
> atomic.weights=c(12.01,1.008,16.00)
> molar.mass(num.atoms,atomic.weights)
[1] 342.296
```

Final Remarks

There are many chemistry, physics, and biology calculations that can be expedited by writing simple R functions. You could say R functions take the chug out of plug and chug.

To this day, my high school slide rule, which belonged to my engineer grandfather, sits atop my desktop computer, beside a FORTRAN punch card, a 5 ¼-inch floppy disk, and a four-function electronic calculator, reminders of the old days when calculation was drudgery, TVs had only three channels, phones were attached to walls, and I had to be bussed for miles through snow-plowed roads to get to school.

WHAT WE LEARNED

1. R has many built-in functions. These functions usually take one or more vectors as input (called the arguments of the function) and perform some calculation with the vectors, returning the result as output.

Example:

```
> x=0:10
> y=sqrt(x)
> x
 [1]  0  1  2  3  4  5  6  7  8  9 10
> y
 [1] 0.000000 1.000000 1.414214 1.732051 2.000000 2.236068
 [7] 2.449490 2.645751 2.828427 3.000000 3.162278
```

2. You can build your own functions in R. You create a function using "dummy" vectors as arguments, which act as place-holders. One or more R statements inside the curly braces of the function define the calculations to be performed with the dummy vectors. The return() statement in the function designates the output of the function. All quantities defined and named within an R function are local to that function and are not available for use outside the function.

Example:

(*run the following script in the R editor*)

```
# Two R functions that convert Fahrenheit temperatures
# to degrees Celsius and vice versa.

degrees.C=function(x) {
  tc=(x-32)*5/9
  return(tc)
}

degrees.F=function(y) {
  tf=9*y/5+32
  return(tf)
}
```

(*type the following commands in the R console*)

```
> oventemps.F=c(325,350,375,400,425,450)
> degrees.C(oventemps.F)
[1] 162.7778 176.6667 190.5556 204.4444 218.3333 232.2222
> oventemps.C=c(160,170,180,190,200,210,220,230)
> degrees.F(oventemps.C)
[1] 320 338 356 374 392 410 428 446
```

Computational Challenges

3.1. Use the molar mass function to calculate the molar masses of the following compounds. You will need to do a bit of research to find the necessary atomic weights.

a. Carbon dioxide
b. Methane
c. Glucose

    d. Iron oxide

    e. Uranium hexafluoride

3.2. Percentage composition. The percentage composition of a chemical compound is the percentage of total mass contributed by each element in the compound:

$$\% \text{ Composition} = \frac{\text{Total mass of element in compound}}{\text{Molar mass of compound}} \times 100.$$

The total mass of an element in a compound is the molar mass of the element multiplied by the number of atoms of that element in the compound. Write an R function to calculate the percentage composition for any element in any compound, given appropriate inputs. Think carefully about the arguments needed and the function's output.

3.3. For each of the three loan equations studied in Chapter 2, write a function to perform the calculations. The equations are reprinted below. Give some careful thought to the design of the input and the output of the functions.

$$\text{Monthly payment} = \frac{(1+s)^m \, sP}{(1+s)^m - 1},$$

$$\text{Principal paid in month } t = \frac{(1+s)^{t-1} \, sP}{(1+s)^m - 1},$$

$$\text{Principal remaining after month } t = P\left[1 - \frac{(1+s)^t - 1}{(1+s)^m - 1}\right].$$

The symbols are as follows: P is the principal amount, m is the total duration of the loan (number of months), s is the monthly interest rate (annual interest rate, expressed as a fraction not a percent, divided by 12), and t is time in months.

3.4. Build yourself a collection of functions to calculate areas, surface areas, and volumes of shapes from geometry. Save them, but do not throw away your geometry text just yet! You are welcome to take advantage of the fact that "pi" in R is a reserved word that returns the value of π to many decimal places (check this out in the console!). In the following table are some shapes and the appropriate formulas.

Symbols in the formulas are as follows. Rectangle: side lengths a, b. Parallelogram: base b, height h. Trapezoid: long base b_1, short base b_2, height h. Triangle: base b, height h. Circle: radius r. Ellipse: long radius r_1, short radius r_2. Rectangular prism: sides a, b, c. Sphere: radius r. Cylinder: base radius r, height h. Cone: base radius r, height h. Pyramid: base side b, height h. Ellipsoid: long radius r_1, first short radius r_2, second short radius r_3.

Areas:	Rectangle	Parallelogram	Trapezoid	Triangle	Circle	Ellipse
	ab	bh	$h(b_1+b_2)/2$	$hb/2$	πr^2	$\pi r_1 r_2$
Surface areas:	Rectangular prism	Sphere	Cylinder	Cone	Pyramid	Ellipsoid
	$2ab + 2ac + 2bc$	$4\pi r^2$	$2\pi r(r+h)$	$\pi r\left(\sqrt{r^2+h^2}+r\right)$	$b\left[2\sqrt{h^2+(b/2)^2}+b\right]$	(No elementary formula)
Volumes:	Rectangular prism	Sphere	Cylinder	Cone	Pyramid	Ellipsoid
	abc	$4\pi r^3/3$	$\pi r^2 h$	$\pi r^2 h/3$	$b^2 h/3$	$4\pi r_1 r_2 r_3/3$

3.5. **Money growth.** Write an R function to calculate the amount of money, n_t, in a fixed interest investment at any time t in the future, starting with n_0 dollars at $100 \cdot i$ % per year interest. The equation is

$$n_t = n_0 \left(1 + i\right)^t.$$

Bonus points awarded if you write the function to take as an argument a whole vector of times and to return a whole vector of corresponding dollar amounts.

3.6. **Hardy–Weinberg frequencies.** Remember from basic genetics that in humans and many animals a gene inherited from the mother and a gene inherited from the father combine to determine a trait inherited by an offspring. If there are two variants of a gene, labeled, say, A and B, in a population, then the possible genotypes of individuals in the population are AA, AB, or BB.

Suppose we model the formation of a new generation as follows. Males and females put all their gametes or sex cells, each of which carries either an A gene or a B gene, into a big urn, and new individuals are made by drawing out two gametes at random. Suppose the proportion (or fraction) of gametes with A genes in the urn was p, and so the proportion of gametes with B genes was $1-p$. These two fractions are called the **gene frequencies** of A and B in the population.

Draw out two gametes at random. The probability of drawing two As and forming an AA genotype individual is p^2, the probability of an AB

individual is $2p(1-p)$ (an AB individual can be formed by drawing AB
or BA), and the probability of a BB individual is $(1-p)^2$. These **genotype
frequencies**, produced theoretically by "random mating," are called the
Hardy–Weinberg frequencies.

Write an R function to take any gene frequency for gene A (where
the frequency is between 0 and 1) and calculate the Hardy–Weinberg
genotype frequencies. The function should take as an argument a single
number and return a vector of three quantities.

3.7. Write yourself a collection of functions to convert length, weight, and volume quantities from English to metric units and from metric to English.
You will need to do a bit of research to find the necessary conversion
constants.

Afternotes: Short Remarks on Topics for Further Study

1. That elements combine in fixed, weight-preserving ratios to form
compounds was recognized more than 200 years ago, by John Dalton
in 1805 (see Patterson 1970), as strong evidence for the existence of
atoms. As additional evidence mounted through the decades, scientists were completely convinced of the existence of atoms, long
before atoms were actually "seen." Twentieth-century chemistry
gave the world wondrous new products, all the while relying on the
theory of atoms. Photographs of uranium atoms made in 1970 by
Albert Crewe with the scanning transmission electron microscope
of his own design (Crewe et al. 1970) came almost as an afterthought.
In science, one does not necessarily have to see something in order to
strongly infer its existence.

2. That certain physical traits are inherited by offspring on average in
predictable, fixed ratios was recognized, by Gregor Mendel in 1866
(see Henig 2000), as strong evidence for a particulate mechanism of
inheritance. We now recognize those particles as genes. That predictable, equilibrium ratios of gene frequencies would result in populations in the absence of forces such as natural selection (the advantage
of one gene over another) or inbreeding was independently derived
in 1908 by Hardy and Weinberg (see Hartl and Clark 2007 for an
elementary derivation). The random mating assumption (along with
some other assumptions such as no natural selection favoring A or B)
behind the Hardy–Weinberg genotype frequencies in a population
might seem unrealistic (surely you are choosey about your romantic
interests?). However, the assumption is tenable with respect to some
traits; for instance, you likely did not consider or even know blood

types when picking your romantic interests. The Hardy–Weinberg frequencies are observed for many traits in many species.

References

Crewe, A. V., J. Wall, and J. Langmore. 1970. Visibility of single atoms. *Science* 168:1338–1340.

Hartl, D., and A. G. Clark. 2007. *Principles of Population Genetics*, 4th edition. Sunderland, MA: Sinauer.

Henig, R. M. 2000. *The Monk in the Garden: The Lost and Found Genius of Gregor Mendel, The Father of Genetics*. Boston: Houghton Mifflin.

Patterson, E. C. 1970. *John Dalton and the Atomic Theory*. Garden City, NY: Anchor.

4

Basic Graphs

It might seem to you that scientists have an obsession with quantification. Your impressions are correct and probably underestimated. To build reliable knowledge, scientists naturally seek to gather and record counts, measurements, and attributes of the items being studied. Data, the collective name for such counts, measurements, and attributes, are the lifeblood of science. Studying data with quantitative tools allows scientists to detect patterns, formulate trial explanations (hypotheses), design tests, make predictions, and assess the reliability of the explanations.

Graphs are crucial to science. Even a small table of data is nearly incomprehensible by itself. Graphs allow scientists to visualize quantitative patterns.

Graphs are a huge part of what R is about. Graphs are R.

Real-World Example

This time, let us go straight to a real-world example. We will explore data from economics and political science. Table 4.1 displays the data that I assembled from the *2010 Economic Report of the President*. The report is available online: http://www.gpoaccess.gov/eop/index.html. Each line of the data represents a federal budget year. Four "variables" or columns are in the data set:

YEAR (1960–2010): I started the data in 1960, as a sort of general beginning to the "modern" government/budget/economic era.

UNEMPLOY: Civilian unemployment rate, in percent.

SURPLUS: Surplus (positive) or deficit (negative) in federal budget, as percent of the gross domestic product that year.

PARTY: Political party of the president who was responsible for that budget year, where R indicates Republican and D indicates Democratic. The federal budget in the first year of a president's term is set by the president in the previous year, and so I used 1 year as a time lag. For instance, 1960 was a Republican year; John F. Kennedy took office in January, 1961, but he had effectively no budgetary influence until the new federal fiscal year began in the following October. Thus, I assigned 1961 as a Republican year as well.

The R Student Companion

TABLE 4.1

Modern U.S. Presidents and Economic Variables

YEAR	UNEMPLOY	SURPLUS	PARTY
1960	5.5	0.1	R
1961	6.7	−0.6	R
1962	5.5	−1.3	D
1963	5.7	−0.8	D
1964	5.2	−0.9	D
1965	4.5	−0.2	D
1966	3.8	−0.5	D
1967	3.8	−1.1	D
1968	3.6	−2.9	D
1969	3.5	0.3	D
1970	4.9	−0.3	R
1971	5.9	−2.1	R
1972	5.6	−2.0	R
1973	4.9	−1.1	R
1974	5.6	−0.4	R
1975	8.5	−3.4	R
1976	7.7	−4.2	R
1977	7.1	−2.7	R
1978	6.1	−2.7	D
1979	5.8	−1.6	D
1980	7.1	−2.7	D
1981	7.6	−2.6	D
1982	9.7	−4.0	R
1983	9.6	−6.0	R
1984	7.5	−4.8	R
1985	7.2	−5.1	R
1986	7.0	−5.0	R
1987	6.2	−3.2	R
1988	5.5	−3.1	R
1989	5.3	−2.8	R
1990	5.6	−3.9	R
1991	6.8	−4.5	R
1992	7.5	−4.7	R
1993	6.9	−3.9	R
1994	6.1	−2.9	D
1995	5.6	−2.2	D
1996	5.4	−1.4	D
1997	4.9	−0.3	D

TABLE 4.1   (*Continued*)

Modern U.S. Presidents and Economic Variables

YEAR	UNEMPLOY	SURPLUS	PARTY
1998	4.5	0.8	D
1999	4.2	1.4	D
2000	4.0	2.4	D
2001	4.7	1.3	D
2002	5.8	−1.5	R
2003	6.0	−3.5	R
2004	5.5	−3.6	R
2005	5.1	−2.6	R
2006	4.6	−1.9	R
2007	4.6	−1.2	R
2008	5.8	−3.2	R
2009	9.3	−10.0	R
2010	9.6	−8.9	D

Source: Data compiled from the 2010 Economic Report of the President (online: http://www.gpoaccess.gov/eop/index.html).

Notes: YEAR, Federal budget year (1960 is Federal fiscal year 10/1959–09/1960, etc.); UNEMPLOY, civilian unemployment (%); SURPLUS, budget surplus or deficit (expressed as % of gross domestic product); PARTY, political party of president responsible for the budget (D, Democratic; R, Republican).

The data provoke curiosity as to whether there are any differences in unemployment and budget surplus/deficit between the budgets of Republican and Democratic presidents. One of the most important things that a president does to influence the economy is prepare/propose/haggle and eventually sign a federal budget. Lots of claims are made by politicians and pundits, but what do the numbers say? It is hard to tell by looking at the table. The data in Table 4.1 practically cry out for visual display.

The data, besides provoking our political curiosity, will serve as a good source of raw material for learning about different types of graphs in R.

We will need each column of the data set to be entered in R as a vector. Take the time now to open the R editor and type the following R statements into a script:

```
year=1960:2010

unemploy=c(5.5, 6.7, 5.5, 5.7, 5.2, 4.5, 3.8, 3.8, 3.6, 3.5,
4.9, 5.9, 5.6, 4.9, 5.6, 8.5, 7.7, 7.1, 6.1, 5.8,
7.1, 7.6, 9.7, 9.6, 7.5, 7.2, 7.0, 6.2, 5.5, 5.3,
5.6, 6.8, 7.5, 6.9, 6.1, 5.6, 5.4, 4.9, 4.5, 4.2,
4.0, 4.7, 5.8, 6.0, 5.5, 5.1, 4.6, 4.6, 5.8, 9.3,
9.6)
```

```
surplus=c(0.1, -0.6, -1.3, -0.8, -0.9, -0.2, -0.5, -1.1, -2.9,
0.3,-0.3, -2.1, -2.0, -1.1, -0.4, -3.4, -4.2, -2.7, -2.7,
-1.6, -2.7, -2.6, -4.0, -6.0, -4.8, -5.1, -5.0, -3.2, -3.1,
-2.8, -3.9, -4.5, -4.7, -3.9, -2.9, -2.2, -1.4, -0.3,  0.8,
1.4, 2.4,   1.3, -1.5, -3.5, -3.6, -2.6, -1.9, -1.2, -3.2,
-10.0, -8.9)

party=c("R", "R", "D", "D", "D", "D", "D", "D", "D", "D",
"R", "R", "R", "R", "R", "R", "R", "R", "D", "D",
"D", "D", "R", "R", "R", "R", "R", "R", "R", "R",
"R", "R", "R", "R", "D", "D", "D", "D", "D", "D",
"D", "D", "R", "R", "R", "R", "R", "R", "R", "R",
"D")
```

Check your typing carefully. Save this script as, say, economics data.R. You will use it repeatedly in this chapter to produce many graphs.

Run the script, and, in the console, test whether the vectors are being stored correctly in R:

```
> year
 [1]  1960 1961 1962 1963 1964 1965 1966 1967 1968 1969 1970
[12]  1971 1972 1973 1974 1975 1976 1977 1978 1979 1980 1981
[23]  1982 1983 1984 1985 1986 1987 1988 1989 1990 1991 1992
[34]  1993 1994 1995 1996 1997 1998 1999 2000 2001 2002 2003
[45]  2004 2005 2006 2007 2008 2009 2010
> unemploy
 [1]  5.5 6.7 5.5 5.7 5.2 4.5 3.8 3.8 3.6 3.5 4.9 5.9 5.6 4.9
[15]  5.6 8.5 7.7 7.1 6.1 5.8 7.1 7.6 9.7 9.6 7.5 7.2 7.0 6.2
[29]  5.5 5.3 5.6 6.8 7.5 6.9 6.1 5.6 5.4 4.9 4.5 4.2 4.0 4.7
[43]  5.8 6.0 5.5 5.1 4.6 4.6 5.8 9.3 9.6
> surplus
 [1]   0.1 -0.6 -1.3 -0.8 -0.9 -0.2 -0.5 -1.1 -2.9  0.3 -0.3
[12]  -2.1 -2.0 -1.1 -0.4 -3.4 -4.2 -2.7 -2.7 -1.6 -2.7 -2.6
[23]  -4.0 -6.0 -4.8 -5.1 -5.0 -3.2 -3.1 -2.8 -3.9 -4.5 -4.7
[34]  -3.9 -2.9 -2.2 -1.4 -0.3  0.8  1.4  2.4  1.3 -1.5 -3.5
[45]  -3.6 -2.6 -1.9 -1.2 -3.2 -10.0 -8.9
> party
 [1]  "R" "R" "D" "D" "D" "D" "D" "D" "D" "D" "R" "R" "R" "R"
[15]  "R" "R" "R" "R" "D" "D" "D" "D" "R" "R" "R" "R" "R" "R"
[29]  "R" "R" "R" "R" "R" "R" "D" "D" "D" "D" "D" "D" "D" "D"
[43]  "R" "R" "R" "R" "R" "R" "R" "R" "D"
```

Note how the vector party is a vector of text characters, rather than numbers. Characters or strings of characters are entered into text vectors with quotations. The symbols R or D represent an "attribute" of the budget year. Such nonnumerical data are called **categorical data** or **attribute data**. Typical attributes recorded in a survey of people are sex, race, political candidate favored, and religion. In many databases, categorical data are coded with numerals (for instance, 0 and 1), but the numerals do not then signify any quantity being measured.

The vectors `unemploy` and `surplus` contain numbers. Such data are called **quantitative data** or **interval data**. The numbers represent amounts of something, and the difference between two data points is the amount by which one exceeds the other. Note: The usual term in science for a column of data values (categorical or quantitative) recorded from subjects is a **variable**. Because variables are stored in R as vectors (text or quantitative), we will often use the words "variable" and "vector" interchangeably when discussing data.

Before we try to graphically contrast the Democratic and Republican differences, we should plot some graphs of each variable alone to obtain an idea of what the data are like.

Make your R editor window active (or your text editor for scripts if you are working in Unix), with the economics data script open. Save the script again with another name, maybe `economics data graphs.R`. We are going to modify this newer script to explore different kinds of graphical displays. The old one will preserve all your typing of the data and can be used as a starting point for future analyses.

Graphs of One Variable

Stripchart

A stripchart (sometimes called a dot plot) is a simple visual portrayal of all the numbers in a quantitative variable. The stripchart consists of a number line, with a symbol, such as a circle, placed on the line at the location of each number.

Add the following command to the end of your script and run the script:

```
stripchart(unemploy,xlab="percent civilian unemployment 1960-
    2010", method="stack",pch=1,cex=3)
```

If you ran the whole script, the vectors of data were read again by R and the existing ones were overwritten; R does not mind the duplication of work. You could have instead just highlighted the new R statement and run it alone because the vectors already had been set up in the workspace. Either way is fine.

Several arguments appear in the stripchart statement. The first argument is the vector of data to use in the chart, in this case `unemploy`. The argument `xlab="Percent civilian unemployment 1960-2010"` produces a descriptive label for the *x*-axis (horizontal axis, which here is just the number line), using the string of characters in the parentheses. The string is chosen by the user. The argument `method="stack"` builds the chart by stacking tied numbers vertically so that they can be seen better. The argument `pch=1` set the plot symbol (plot character) code to be 1, which is the R code for a circle. I like circles better than the squares that R uses as a default. Finally, `cex=3` enlarges the circles to three times the default size, making the graph somewhat better for viewing on a projector from the back of a room or for the severe reduction that occurs with publication.

The last three arguments in the `stripchart()` function were just my preferences. R is flexible! Some of the frequently used arguments appear in Chapter 13, and more are catalogued in Appendix C.

But first, inspect the graph itself (Figure 4.1a). A column of raw numbers comes to life when we can "see" them. We see at a glance the range of values of unemployment that the nation has faced since 1960. A small number of years had unemployment above 9%, very severe. The better years featured unemployment below 4%. The bulk of the years had unemployment between 4% and 8%.

Histogram

Different types of graphs emphasize different aspects of the data for study. Perhaps, we might want a better visual representation for comparing the frequencies of years with unemployment above 9%, below 4%, between 5% and 6%, and so on. A histogram is one such graphical tool. A histogram represents the frequency of data points in an interval as a rectangle over the interval, with the area of the rectangle equal to the frequency.

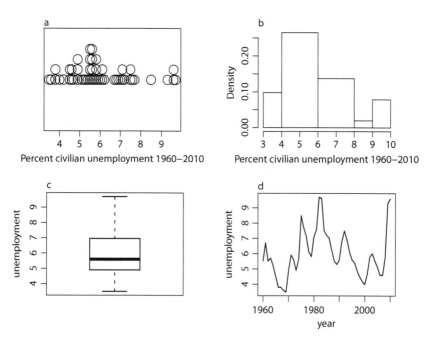

FIGURE 4.1
Graphical displays of U.S. civilian unemployment, 1960–2010. (a) Stripchart (or dot plot). (b) Histogram. (c) Boxplot. (d) Timeplot.

Let us try making a histogram. In your script, "comment out" the strip-chart command, and add a histogram command:

```
# stripchart(unemploy,xlab="Percent civilian unemployment
# 1960-2010", method="stack",pch=1,cex=3)
hist(unemploy, main=" ",xlab="Percent civilian unemployment
  1960-2010")
```

Run the script, and a histogram should appear. Before interpreting the graph, let us examine the arguments used in the `hist()` statement. The `main=" "` argument suppresses the printing of a main title above the graph (or more precisely, prints nothing but a blank space). If you would like a title, put the text characters for it between those quotes. The other arguments are the same as in the `stripchart()` command. Many of the arguments work the same in different graphical commands.

The histogram statement automatically picks reasonably nice intervals for representing the data. Here, it picked the intervals 3 to 4, 4 to 5, 5 to 6, and so on. Any observation on a boundary between two intervals is counted in the lower interval. The user can provide alternative intervals if desired. For instance, if the desired end points of the intervals are 3, 4, 6, 8, 9, 10, then put the optional argument `breaks=c(3,4,6,8,9,10)` in the histogram statement. Here, `c(3,4,6,8,9,10)` produces a vector with the desired end points as elements. The histogram statement would look like this:

```
hist(unemploy,main=" ",breaks=c(3,4,6,8,9,10),xlab="Percent
civilian unemployment 1960-2009")
```

Close the graph window and rerun the script with these new intervals for the histogram. The resulting plot (Figure 4.1b) illustrates something very important about histograms: the information about frequency is carried by the *areas* of the rectangles, not the heights. A histogram is not a bar graph. The new end points created intervals with unequal widths. The fraction or proportion of the unemployment numbers in each interval is the width of the interval times the height.

Think of each rectangle as a cake observed from above, with the amount of cake over the interval equal to the fraction or proportion of data points in the interval. The longest cake might not necessarily represent the biggest cake, except of course when the cake widths are all equal.

Stem-and-Leaf Plot

A stem-and-leaf plot is a clever arrangement of the actual numbers in the variable into a sort of histogram. Close the graph window, comment out the histogram statement in your R script, and try the simple command for a stem-and-leaf plot:

```
# stripchart(unemploy,xlab="Percent civilian unemployment
# 1960-2009", method="stack",pch=1,cex=3)
# hist(unemploy, main=" ",breaks=c(3,4,6,8,9,10),xlab="Percent
# civilian unemployment 1960-2009")
stem(unemploy)
```

This time, the output appears in the console:

```
The decimal point is at the |
  3 | 5688
  4 | 0255667999
  5 | 12345555666678889
  6 | 0112789
  7 | 01125567
  8 | 5
  9 | 367
```

A stem-and-leaf plot is a typographic construct that allows the viewer to reconstruct all the original data, to at least two digits. Each data point has a stem and a leaf. The stem is a number in the left column of the plot, and the leaves are the numbers extending horizontally to the right. For a data point in the variable unemploy, the stem gives the digit to the left of the decimal (the ones digit), and the digit to the right of the decimal is found as one of the leaves. Look at the stem-and-leaf plot and see if you can tell that the smallest unemployment years featured 3.5%, 3.6%, 3.8%, and 3.8% unemployment.

Stem-and-leaf plots have varying designs, depending on the range and quantity of data in the variable. The stems and the leaves that R picks usually result in decent plots.

Boxplot

A boxplot (often called a box-and-whiskers plot) relies on five numbers to summarize all the data in a variable. The first two numbers are the minimum and the maximum, easy enough to understand. The third number is the median of the data. The median is one way to define the "middle" of the data, and it is calculated as follows. (1) Order the data from smallest to largest. (2) If the number of data points is an odd number, the median is the data point in the middle having an equal number of smaller and larger data points; or if the number of data points is even, the median is the average of the two middlemost data points. Here are a few examples:

23	29	36	38	42	47	54		7 observations: odd
median is 38
23	29	36	38	43	47	54	55	8 observations: even
median is (38 + 43)/2 = 40.5 |

The median is sometimes called the 50th percentile because approximately 50% of the data points are below the median and approximately 50% are above. The median is not the same as the mean (or arithmetic average).

Of course R has functions for calculating the minimum, maximum, and median! Try them in the console:

```
> min(unemploy)
[1] 3.5
> max(unemploy)
[1] 9.7
> median(unemploy)
[1] 5.6
```

We already can tell that these three numbers give us some sense of what the data are like. Unemployment ranged from 3.5% to 9.7%, with unemployment in about half of the years above and about half below 5.6%.

The final two numbers used in a boxplot are simple: the median of all the data points that are less than the median, and the median of all the data points that are greater than the median. Those two numbers are called the 25th percentile and the 75th percentile—the first is greater than approximately 25% of the data, while the second is greater than approximately 75% of the data. The term percentile is no doubt familiar (perhaps painfully) to contemporary students reared on a stark educational diet of high-stakes standardized testing.

So, the so-called five-number summary that goes into a boxplot consists of: minimum, 25th percentile, 50th percentile (median), 75th percentile, maximum. Close any open graphs, go to your script now, comment out any unneeded commands, and make a boxplot to view these numbers! The R command is as follows:

```
boxplot(unemploy)
```

Or if you wish, put in a descriptive label for the *y*-axis (vertical axis):

```
boxplot(unemploy,ylab="Percent civilian unemployment 1960-2010")
```

In the resulting boxplot (Figure 4.1c), the ends of the box are the 25th and the 75th percentiles, the ends of the "whiskers" locate the minimum and the maximum, and the median is shown with a line. One can see a departure from symmetry in that the median unemployment is closer to the low end than the high end. Asymmetry of the spread of lower and higher values in data is called **skewness**.

Timeplot

The variable `unemploy` is what is called a time series: its values are recorded through time. The variable `surplus` is a time series too. A timeplot is a simple *x-y* graph of a time series variable (usually the vertical axis) and time (the horizontal axis). It is useful for studying any trends or patterns in a time series through time. A timeplot is technically a graph of two variables: the time series and the time. However, the interest is not necessarily in how the variable time changes; rather, the focus is on the time series variable. I think of the timeplot as a graph of one variable.

A timeplot is produced by graphing the variable values and their corresponding times as points (ordered pairs) on an *x-y* graph. A key decision is what symbols to use for the points. I prefer to simply connect the points with lines, but not to use any symbols for the points themselves, resulting in a clean-looking graph uncluttered with distractions. Perhaps, I might use point symbols, such as circles and squares, in order to distinguish two time series plotted on the same chart. In this sense, using a point symbol for just one time series is like writing an outline in which there is a Part A but not a Part B.

We know something about *x-y* plots already, having seen some in Chapters 1 and 2. Try a timeplot of unemploy. The drill should be getting familiar by now: close any open graph windows, and in your script, comment out unwanted commands, and add the following command for a timeplot:

```
plot(year,unemploy,type="l",xlab="Year",ylab="Civilian
    unemployment")
```

Run the script, and the timeplot should appear (Figure 4.1d). The graph shows peaks and valleys of unemployment, with two major episodes of unemployment being the early 1980s and the last years in the data (2008–2010).

Your script now has accumulated some handy graphing commands! You can use it in the future and put in other data, comment out, un-comment, and modify the lines of R statements as needed. Save your script and we will move on to more graphs. Know that the first computing challenge at the end of the chapter is to produce all the above graphs using the other economic variable, surplus. If that does not actually sound very challenging now, you are becoming an accomplished R user!

Graphs of Two Variables

Various types of graphs are helpful for depicting relationships between two variables.

If you like, you can use your script, economic data plots.R, to hold the following R statements for two-variable plots as well as the single-variable plots. Or if you prefer, you can dedicate a new script; save the script under a new name, perhaps economic data bivariate plots.R, and erase all the one-variable plot commands (but not the commands that set up the vectors from the data). Either way is fine and is a matter of personal programming style (are you a lumper or a splitter?). I tend to be a splitter. Files are cheap; debugging time is expensive.

Scatterplot

A scatterplot can be used when both variables of interest are quantitative. A scatterplot portrays the values of two variables recorded from each subject or item as an ordered pair on an *x-y* plot. We have seen a scatterplot of the wolf–moose data in Chapter 1 (Figure 1.2). Our example scatterplot here will use the values of unemploy and surplus recorded in each year as the ordered pairs. Add the plot command to your script. Your script will look like this, with any other graphing commands erased or commented out:

```
year=1960:2010

unemploy=c(5.5, 6.7, 5.5, 5.7, 5.2, 4.5, 3.8, 3.8, 3.6, 3.5,
4.9, 5.9, 5.6, 4.9, 5.6, 8.5, 7.7, 7.1, 6.1, 5.8,
7.1, 7.6, 9.7, 9.6, 7.5, 7.2, 7.0, 6.2, 5.5, 5.3,
5.6, 6.8, 7.5, 6.9, 6.1, 5.6, 5.4, 4.9, 4.5, 4.2,
4.0, 4.7, 5.8, 6.0, 5.5, 5.1, 4.6, 4.6, 5.8, 9.3,
9.6)

surplus=c(0.1, -0.6, -1.3, -0.8, -0.9, -0.2, -0.5, -1.1, -2.9,
0.3, -0.3, -2.1, -2.0, -1.1, -0.4, -3.4, -4.2, -2.7, -2.7,
-1.6, -2.7, -2.6, -4.0, -6.0, -4.8, -5.1, -5.0, -3.2, -3.1,
-2.8, -3.9, -4.5, -4.7, -3.9, -2.9, -2.2, -1.4, -0.3, 0.8,
1.4, 2.4, 1.3, -1.5, -3.5, -3.6, -2.6, -1.9, -1.2, -3.2,
-10.0, -8.9)

party=c("R", "R", "D", "D", "D", "D", "D", "D", "D", "D",
"R", "R", "R", "R", "R", "R", "R", "R", "D", "D",
"D", "D", "R", "R", "R", "R", "R", "R", "R", "R",
"R", "R", "R", "R", "D", "D", "D", "D", "D", "D",
"D", "D", "R", "R", "R", "R", "R", "R", "R", "R",
"D")

plot(surplus,unemploy,type="p",ylab="unemployment")
```

Now run the script. The resulting graph reveals a strong negative relationship between surplus and unemploy (Figure 4.2a). Years with high surplus are associated with years of low unemployment.

Be careful about jumping to conclusions. Such an association is not necessarily causal. Both variables might be changing in response to other "lurking" variables, or there might be important time lags in the responsiveness of economic variables to government policies. Try formulating several hypotheses that would explain the pattern. Building multiple working hypotheses is a critical mental habit of rigorously trained scientists. The habit helps an

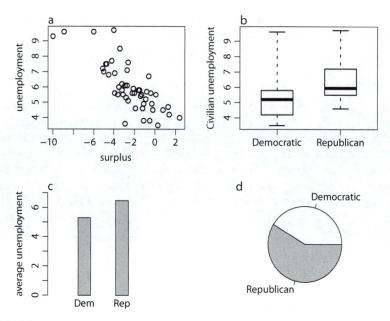

FIGURE 4.2
Graphical displays of U.S. economics data, 1960–2010. (a) Scatterplot of budget surplus
(horizontal axis) and civilian unemployment (vertical axis). (b) Side-by-side boxplots of civilian unemployment under Democratic and Republican presidential budget years. (c) Bar
graph of mean unemployment during Democratic and Republican presidential budget years.
(d) Pie chart of number of budget years under Democratic and Republican presidents.

investigator avoid becoming too attached to or invested in one explanation, a
predicament that famously can cause investigators to ignore strong evidence
against their pet explanations.

That being said, scatterplots are a warhorse of science. Scatterplots and
other two-variable graphs are abundant in the pages of original scientific
reports.

Side-by-Side Boxplots

Here, at last, we can look at any association between a president's party affiliation and economic variables. A nice technique for comparing the values of a
quantitative variable across the categories of a categorical variable is to draw
side-by-side boxplots. Each boxplot contains only the numbers in the quantitative variable that correspond to one particular category in the categorical
variable. The following R statement will produce side-by-side boxplots of
unemploy, separated according to the categorical variable party:

```
boxplot(unemploy~party, range=0, names=c("Democratic",
   "Republican"), boxwex=.5, ylab="Civilian unemployment")
```

Comment out the scatterplot() statement, type in the boxplot() statement, and run the script. The resulting picture (Figure 4.2b) portrays a fairly evident difference between Democratic president budget years and Republican president budget years with regard to unemployment. Certainly, there is an overlap of the two groups, but the Democratic unemployments as a whole tend to have lower values than the Republican unemployments.

The graph shows an association between a quantitative variable (unemployment) and a categorical variable (party). If there were little or no association, the five points of each boxplot would be about the same height. Again, an association is not necessarily a causal relationship. Are there potential lurking variables? What if the time lag between budget policy and unemployment response was a year or more?

The arguments of the boxplot statement here are as follows. The key argument is unemploy~party. This is a frequent syntax in R arguments, and it means, loosely, "unemployment by party." The result is that the boxplot function produces plots separated by the levels of party. If party had more than two levels, then there would be more than two boxplots produced.

The range=0 argument extends the whiskers to the maximum and the minimum values (the ordinary boxplot definition). Occasionally, data sets have "outliers" or points so extreme that they might have been caused by some mechanism other than that which produced the rest of the original data. Variants of boxplots extend the whiskers just to the limits defined by outlier detection criteria, with suspected outliers flagged as separate symbols outside the whiskers.

The names=c("Democratic","Republican") argument sets the printed names of the categories in the categorical variable to be the character strings in the vector produced by c("Democratic","Republican"). How did we know Democratic values would be plotted on the left of the graph and Republican on the right, assuming R has no knowledge of political leanings? R simply takes the codes used in the categorical variable and prints the results on the graph according to sorted or alphabetical order.

The boxwex=.5 argument sets the width of the boxes to half (.5) of the normal default width. I like narrower boxes when comparing two or more side by side.

Bar Graphs and Pie Charts

Bar graphs and pie charts are available in R. Bar graphs and pie charts are popular in business and management, but they are not used quite as much in the natural sciences. Many times, instead of bar graphs or pie charts, one can find more informative ways of depicting the data.

Bar graphs and pie charts are simple in R. The commands for producing these only need, as a minimum, some vector of numbers as an argument. Try the following commands at the console:

```
> x=c(4,5,7,9)
> barplot(x)
```

A rather basic, stripped-down bar graph of the numbers in x results. We will put in some options to make things look better below. Now close the graph window and type another command at the console:

```
> pie(x)
```

A simple pie chart appears. You will see the advent of color for the first time in the activities in this book; R picks some eye-friendly pastel colors for the pie wedges. R has vast color resources for its graphics. However, to keep the cost of this book as low as possible, all graphics in this book are printed in grayscale.

Bar Graphs and Pie Charts Using Data

Let us try a bar graph for some aspects of our data. Suppose we want to draw the mean (or average) unemployments of Democratic and Republican budgets side by side as bars. The barplot() command in R for bar graphs will want a vector of quantities to draw as bars. We will have to calculate the two means and put them into a vector for plotting. We will take a little time here to explore some ways in R of extracting particular things in data that we might wish to portray graphically.

Somehow, we have to separate out all the Democratic and all the Republican unemployments from unemploy. We could just do this by inspection and build two vectors with two c() commands. However, there are some powerful and clever ways of picking elements out of vectors in R. Going through them all here would make this chapter too long, but it will be fun to introduce one of them here and save us some typing. After we separate out Democratic and Republican unemployments, we can use the mean() function that we saw in Chapter 3 for calculating the means for graphing.

Go to the console and type:

```
> zz=c(1,5,-4,2)
> aa=c(TRUE,FALSE,TRUE,FALSE)
> zz
[1]  1  5 -4  2
> aa
[1]  TRUE FALSE TRUE FALSE
```

Make sure that TRUE and FALSE are in upper case. Here, zz is just an ordinary vector with a few numbers as elements. The vector aa, though, is a special type of vector in R, called a logical vector. Logical vectors can be created with various R commands that perform comparisons of elements in two different vectors. More importantly for us, logical vectors can be used to pick elements out of vectors. Type (note the use of square brackets):

```
> zz[aa]
[1]  1 -4
```

You can see that zz[aa] is a vector that contains the elements of zz corresponding to the TRUE elements in aa. We can use this device to separate the Democratic and the Republican unemployments.

The vectors unemploy and party should be in your workspace waiting to be used. At the console, type (note the use of two equals signs):

```
> party=="R"
 [1]  TRUE  TRUE FALSE FALSE FALSE FALSE FALSE FALSE FALSE FALSE
[11]  TRUE  TRUE  TRUE  TRUE  TRUE  TRUE  TRUE  TRUE FALSE FALSE
[21] FALSE FALSE  TRUE  TRUE  TRUE  TRUE  TRUE  TRUE  TRUE  TRUE
[31]  TRUE  TRUE  TRUE  TRUE FALSE FALSE FALSE FALSE FALSE FALSE
[41] FALSE FALSE  TRUE  TRUE  TRUE  TRUE  TRUE  TRUE  TRUE  TRUE
[51] FALSE
```

The "two equals" (==) is a logical comparison operator. It compares every element of party to the text character R and produces a logical vector in response containing TRUE where there are Rs and FALSE where there are Ds (or more precisely, anything else besides Rs). That looks like what we need! Let us try it first in the console before adding it to the script. Type:

```
> unemploy[party=="D"]
 [1]  5.5 5.7 5.2 4.5 3.8 3.8 3.6 3.5 6.1 5.8 7.1 7.6 6.1 5.6
[15]  5.4 4.9 4.5 4.2 4.0 4.7 9.6
```

An inspection of Table 4.1 confirms that the statement unemploy [party=="D"] picked out all the Democratic unemployments. Now try the following in the console:

```
> mns=c(mean(unemploy[party=="D"]),mean(unemploy[party=="R"]))
> mns
[1]  5.295238 6.463333
```

The vector mns contains the two means (Democtatic and Republican) we were after! Let us go through the statement mns=... in detail. In that statement, arguments are nested inside arguments, making it look complicated, but you will see that it is simple when we start from the inside and work outwards. Inside, party=="D" is the logical vector containing TRUE for Democratic and FALSE for Republican budget years. Then, unemploy[party=="D"] is the vector we saw before with all the Democratic unemployments picked out of unemploy. The Republican unemployments likewise are picked out with unemploy[party=="R"]. Next, mean(unemploy[party=="D"]) just calculates the mean of those Democratic unemployments. The Republican mean is produced similarly. Finally, the c(mean(unemploy[party=="D"]),mean (un-employ[party=="R"])) statement builds a vector out of the Democratic and the Republican means.

We could have written this differently, with more but smaller statements. For instance,

```
> dd=party=="D"
> rr=party=="R"
> unemp.D=unemploy[dd]
> unemp.R=unemploy[rr]
> D.mean=mean(unemp.D)
> R.mean=mean(unemp.R)
> mns=c(D.mean,R.mean)
> mns
[1] 5.295238 6.463333
```

Can you follow that more easily? Likely so, if you just go through the statements one by one. With R, sometimes just spreading calculations over many statements and naming/storing everything as you go is clearer. Storage for our purposes is only limited by our ability to invent new names for the vectors. Now, there is a cost of long scripts: more statements make R run slower. In other words, R runs faster if the same calculations are concentrated into fewer statements. However, R is so fast that only high-end power users of R (who might be conducting big simulations that require days to run!) will notice much difference.

Long or concentrated—either way works. Many ways work in R. Perhaps the best advice for a complex project is get your calculations to work in long form at first and only then season your scripts to taste with pure cool concentrated cleverness.

Oh yes, bar graphs! We wanted these means for a bar graph. I think we have all the necessary pieces in place.

Go back to the script, comment out all the previous graph commands, and add:

```
mns=c(mean(unemploy[party=="D"]),mean(unemploy[party=="R"]))
barplot(mns)
```

Run the script. A rather stark-looking bar graph appears. Would you like thinner bars, some labels, more spacing between bars, and a longer vertical axis with a label? In the script, revise the barplot() command by inserting some options:

```
barplot(mns,xlim=c(0,1),width=.1,ylim=c(0,7),
  ylab="average unemployment",names.arg=c("Dem","Rep"),space=2)
```

Run the script now, and the bar graph should look a bit nicer (Figure 4.2c). The options specified here are as follows. The xlim=c(0,1) defines an x-axis to start at 0 and end at 1. Of course, there is no quantity being portrayed on

the *x*-axis. Rather, the `xlim` argument just gives a scale for the specification of
the widths of the bars. The widths then are given by the `width=.1` argument,
setting each bar to have a width of 1/10. The `ylim=c(0,7)` argument defines
a *y*-axis to start at 0 and end at 7, so that each bar can be compared to the scale.
The default *y*-axis for our graph was too short. The `ylab="average unem-`
`ployment"` provides the text string for labeling the vertical axis. The `names`
`.arg=c("Dem","Rep")` argument provides the text strings for labeling the
bars. Finally, the `space=2` argument sets the spacing between bars to be 2
bar widths (you can play with this setting until the graph looks nice to you).

The `pie()` command for making pie charts takes a vector as an argument
containing the numbers to be compared. The relative sizes of the numbers
are represented in the picture as pie slices. Pie charts compare quantities as
proportions of some total amount. In our economy data, perhaps, we could
look at the proportion of years with Democratic budgets versus the proportion of Republican years.

We can use the logical device from the above barplot statements to pick
out the Democratic and the Republican years, only this time we will simply
count them rather than calculate means as we did for unemployment. Recall
from Chapter 3 that the `length()` function counts the number of elements
in a vector. Are we feeling confident? Let us "wing it" and write some trial
statements in our script:

```
num.yrs=c(length(year[party=="D"]),length(year[party=="R"]))
pie(num.yrs)
```

I do not know about you, but instead of typing, I just copied the two barplot statements from before, pasted them, and then changed them into the
pie chart statement (changing "mean" to "length" etc.). Once you start accumulating functioning scripts for various tasks, working in R amounts to a lot
of copying and pasting. Comment out all the previous graphing commands
and run the script.

OK, the resulting picture is a bit uninformative; it could use some descriptive labels. In the `pie()` command in the script, add a `labels=` argument:

```
pie(num.yrs,labels=c("Democratic","Republican"))
```

Close the window with the graph and rerun the script. Better? We see (Figure
4.2d) that since 1960, Republican presidents have been responsible for somewhat more budget years than Democratic presidents.

Final Remarks

One lingering inconvenience seen in the above scripts is data management.
Having to type something like `xamounts=c(29,94,...)` for each variable in
a large data set with many variables would get very old very fast. Bringing

files of data in and out of R is a cornerstone of the convenience of using R. Let us take up data management right away in Chapter 5.

Contemporary graphical techniques in the sciences feature many ways of depicting relationships between variables. No book about R would be complete without exploring some of the advanced graphical capabilities of R. In this book, Chapter 13 serves this purpose.

In this chapter on basic graphs, we avoided use of color. The pie charts we produced automatically shaded the slices with some default colors, but otherwise, all our graphs were black and white. Color can provide a tremendous enhancement to graphs. However, color also can become grossly distracting when misused, and misuse is unfortunately more the rule rather than the exception in the graphics served up by media in our daily lives. I regard use of color as an advanced graphical tool, mostly for studying after this book is mastered. We will postpone further consideration of color until Chapter 13, and even there, the discussion will be brief.

Appendix C gives a systematic list of the most-used options available in the R graphics functions appearing in this book.

WHAT WE LEARNED

1. Graphical functions in R usually take vectors of data (called "variables") as arguments and produce various types of graphical displays. Optional arguments in the graphical functions provide customizing features such as descriptive axis labels.
2. Variables are "quantitative" if they contain numerical data (such as height or weight) and are "categorical" if they contain codes for categories or attributes (such as sex or political affiliation).

Examples:

```
> heights=c(65,68,72,61)  # quantitative variable
> sex=c("F","M","M","F")  # categorical variable
> heights
[1] 65 68 72 61
> sex
[1] "F" "M" "M" "F"
```

3. A histogram, stem-and-leaf plot, boxplot, and timeplot are types of graphical displays for variables (vectors) containing quantitative data.

Examples:

```
#  In the following R script, un-comment the statement
#  corresponding to the desired graph.
```

```
#----------------------------------------------------
quiz.scores=c(74,61,95,87,76,83,90,80,77,91)
# histogram(quiz.scores)
# stemplot(quiz.scores)
# boxplot(quiz.scores)

# Script for a time plot.
#----------------------------------------------------
dry.weight=c(0.10,0.17,0.27,0.36,0.42,0.46,0.48,0.49,0.49)
time=0:(length(dry.weight)-1)
plot(time,dry.weight,type="l")
```

4. A scatterplot and side-by-side boxplot are useful for depicting associations between two variables.

   Examples:

```
# Scatterplot example script.
# Data from the election-at-large for city council,
# Moscow, Idaho, 1999. Seven candidates ran, highest
# number of votes got a city council seat.
# Variables are dollars spent by candidates on their
# campaigns, and the resulting number of votes garnered.
#----------------------------------------------------
dollars.spent=c(0,0,404,338,583,1992,1849)
number.votes=c(159,305,706,912,1159,1228,1322)
plot(dollars.spent,number.votes,xlab="Dollars spent",
  ylab="Number of votes")


# Side-by-side boxplot example script.
# Data from Margolese 1970 Hormones and Behavior.
# Androsterone levels (mg in 24 hr in urine sample)
# and sexual orientation (S: straight, G: gay)
# in 26 human males.
#----------------------------------------------------
level.andro=c(3.9,4.0,3.8,3.9,2.9,3.2,4.6,4.3,3.1,2.7,
  2,3,2.5,1.6,3.9,3.4,2.3,1.6,2.5,3.4,1.6,4.3,2.0,
  1.8,2.2,3.1,1.3)
orientation=c("S","S","S","S","S","S","S","S","S","S",
  "S","G","G","G","G","G","G","G","G","G","G","G","G",
  "G","G","G")
boxplot(level.andro~orientation, range=0, names=
  c("Gay","Straight"),boxwex=.5,
  ylab="Androsteronelevel")
```

5. Logical vectors contain the elements TRUE and FALSE, all in upper case, with no quote marks. Logical vectors can be used to pick elements out of other vectors.

Example:

```
> x=c(4,-1,0,12,19,-2)         # x is an ordinary vector
> p=c(TRUE,TRUE,FALSE,TRUE,TRUE,FALSE)  # p is a
  logical vector
> y=x[p]
> y
[1]   4 -1 12 19
```

6. Bar graphs and pie charts are available in R.

Examples:

```
# Bar graph example script.
# Percent of young voters, ages 18-29, voting
# Democratic in US presidential elections,
# 1992-2008. Source: Pew Research.
Young.percent= c(43,53,48,54,66)
barplot(young.percent,xlim=c(0,1),width=.05,ylim=c(0,70),
    names.arg=c("1992","1996","2000","2004","2008"))

# Pie chart example script.
# US Federal Government spending, in billions $,
# fiscal year 2010. Source: Congressional Budget
# Office.
spending=c(416,197,793,701,689,660)
pie(spending,labels=c("other mandatory","net interest",
    "medicare & medicaid","social security",
    "defense department","discretionary"))
```

Computational Challenges

4.1. Obtain the various one-variable graphs for the variable `surplus` in the economics data in Table 4.1. Be sure to relabel everything in the graphs appropriately.

4.2. Obtain the side-by-side boxplot and the bar graph of the means, comparing Democratic and Republican presidential budget years, for the variable `surplus` in the economics data in Table 4.1. Be sure to relabel everything in the graphs appropriately.

4.3. Collect data from your class: height, sex, age, length of ring finger divided by length of middle finger (decide on a standard method of measuring these finger lengths), number of text messages sent that day, and any other quantitative or categorical variables that might be of interest. The variables should be nonintrusive, in good taste, fun for all, and provoking curiosity. Enter the data as variables in R. (A) Characterize the quantitative variables with various one-variable graphs. (B) Explore possible relationships with various two-variable graphs. (C) Write at least two plausible (at first, anyway) hypotheses explaining any relationships that appear.

4.4. For the following chemical compounds, draw bar graphs or pie charts depicting the percent composition by molar mass of each chemical element in the compound. You will need to do a bit of Internet or textbook research to find the chemical compositions of the compounds. The function for percent composition you constructed as a computational challenge in Chapter 3 might come in handy.

  A. Adenosine

  B. Guanine

  C. Cytosine

  D. Thymine

Afternotes

1. Many web sites that provide instruction in R graphics and examples of different types of graphs executed in R exist. A few are listed in the following. Do not miss, for instance, the fabulous "addictedtor" site that showcases spectacular real graphics produced in R for scientific publications.

   http://www.harding.edu/fmccown/r/

   http://www.statmethods.net/graphs/

   http://addictedtor.free.fr/graphiques/

   http://www.ats.ucla.edu/stat/r/library/lecture_graphing_r.htm

2. How does science make conclusions about causal relationships? In situations in which experiments are possible, one can manipulate a quantitative variable and observe the response of the other, holding the values of other possible explanatory variables constant. A scatterplot of the manipulated variable and a variable hypothesized to change in response to the manipulated variable can provide a conclusive visual argument for a causal relationship.

   In situations in which experiments are not possible, strong evidence can emerge when an explanation accounts for an interconnected web of associations of many variables, with parts of the web strongly contradicting other explanations. For instance, epidemiologists established that smoking caused cancer in humans without conducting the experiment on humans. The collective body of evidence for that conclusion is vast, with original reports spanning many hundreds of scientific journals.

5

Data Input and Output

The data in Table 5.1 are from 90 students enrolled recently in an introductory statistics course at a state university. The students were from many different majors across sciences, social sciences, and humanities. The course was listed as a sophomore-level course, and most of the students had indeed already completed at least a year of college. Each row of the data represents one student.

The main variable of interest recorded from each student was the student's current cumulative grade point average (GPA) in college courses completed at the university. The GPA was to be used as an index of college success. Other variables were recorded as potential predictors of GPA and included composite ACT score (if the student took the SAT instead, the combined SAT score was transformed into the equivalent percentile ACT score), high school GPA, sex (male or female), and type of residence ("Greek" fraternity or sorority, university residence hall, or off-campus).

The data naturally provoke some curiosity among college age and precollege age students. The high-stakes national tests, high school grades, and college performance are sources of great stress. For us here, now that we know a bit about R, the data provoke an overwhelming desire to draw some graphs!

Except for all that typing. Ninety lines of data, yuck. Maybe graphs can wait until some later day.

Data Frames in R

Happily, the R people have created solutions for almost every laborious impediment to good quantitative scientific analysis. Data input and output in R is easy and elegant.

A data set entered into R that is ready for analysis is called a **data frame**. A data frame is basically a list of vectors. You can think of a data frame as a rectangular table of data, with the columns being vectors of numerical or categorical data. The rows of the data frame correspond to the individuals or subjects from which the data variables were recorded. In our example here, each row in the data frame we will build represents a student, but in other applications, the subjects of the study might be potted plants, towns, cancer patients, laboratory mice, or computer memory chips.

TABLE 5.1

Observations on College GPA (univGPA), Composite ACT Score (ACT), High School GPA (hsGPA), Sex, and Housing Type for 90 Students in a University Introductory Statistics Course

UnivGPA	ACT	HsGPA	Sex	Housing
3.40	24	3.73	m	o
3.25	30	3.43	m	o
3.47	24	3.78	m	o
3.63	24	3.00	m	o
1.80	27	3.20	m	o
3.60	19	3.30	m	r
3.70	26	4.00	m	g
3.80	23	3.30	m	g
2.70	22	3.30	m	o
3.43	24	3.60	m	g
3.00	14	3.00	f	o
3.90	32	3.89	f	r
3.52	25	3.92	f	g
3.67	28	3.80	m	o
3.45	24	2.30	m	o
3.06	26	3.60	m	r
2.80	21	3.40	m	r
3.25	22	3.80	f	r
3.00	22	3.40	m	g
3.40	26	3.60	m	o
3.30	25	3.35	m	o
3.40	24	3.70	f	o
3.02	27	3.97	m	o
2.60	24	3.35	m	o
3.20	17	3.50	m	o
4.00	29	4.00	f	g
2.15	31	2.87	m	o
3.80	25	4.00	f	g
3.67	27	4.00	f	g
3.40	24	3.80	f	o
4.00	25	4.00	f	o
2.89	24	3.69	m	o
3.00	28	3.80	m	o
4.00	23	3.50	f	o
3.70	26	3.00	f	o
4.00	28	4.00	f	r
2.80	22	3.85	m	g
3.80	18	3.00	m	o
2.80	18	3.00	m	r
2.10	28	2.50	m	o
3.00	29	3.80	f	r
3.70	22	3.87	m	g
3.90	26	4.00	m	g
3.70	29	4.00	f	g
3.40	22	3.75	f	g
3.70	30	3.90	m	r
3.00	26	3.65	f	o

TABLE 5.1 *(Continued)*

Observations on College GPA (univGPA), Composite ACT Score (ACT), High School GPA (hsGPA), Sex, and Housing Type for 90 Students in a University Introductory Statistics Course

UnivGPA	ACT	HsGPA	Sex	Housing
3.80	23	4.00	f	o
3.87	30	4.00	f	g
3.14	20	2.41	m	g
3.51	21	3.20	m	o
3.72	25	3.76	m	o
2.91	24	3.25	m	o
3.49	26	4.00	f	r
3.20	30	3.40	f	g
3.60	28	3.85	f	g
3.50	25	3.90	f	r
3.30	32	3.70	f	g
2.48	23	3.54	f	o
3.30	27	3.70	f	g
3.20	29	3.60	f	g
3.75	29	4.00	f	g
3.75	27	3.90	f	o
2.80	17	3.20	m	o
3.70	29	3.50	f	o
4.00	35	3.90	m	o
3.20	20	3.10	f	o
3.00	32	3.40	m	o
4.00	31	4.00	m	o
3.05	24	3.75	f	g
4.00	28	3.90	f	g
3.00	32	3.80	m	o
3.31	26	3.75	m	o
3.00	24	3.60	f	o
3.23	28	4.00	m	o
3.00	27	3.75	f	o
2.15	19	3.30	m	o
3.20	21	2.50	m	o
3.20	21	3.60	m	o
3.00	30	2.98	m	o
3.20	32	3.50	m	o
3.71	26	3.81	f	g
3.68	32	3.93	f	g
3.00	23	3.98	f	o
2.50	18	2.26	m	o
3.40	24	2.40	m	g
4.00	28	4.00	f	g
2.12	19	3.90	m	r
3.72	31	3.70	f	r
2.05	19	2.90		

Notes: o, off campus; r, on-campus; g, sorority/fraternity.

The basic idea is for the data to be stored on the computer system in the form of an ordinary file, usually a text file. Inside R, there are commands to read the file and set up a data frame in the R workspace. Vectors representing the columns of data are then available in the workspace for whatever R analyses are desired. Additional commands exist to arrange the results of the analyses into a data frame and to output the resulting table in the form of a file stored on the computer system.

Of course, there is no avoiding the original typing of the data into the computer file. However, once the data are stored, they can be analyzed in many ways with many different R scripts. As well, data from previous studies by others frequently exist somewhere in file form. For instance, the college GPA data in Table 5.1 are posted online in this book's list of data files: http://webpages.uidaho.edu/~brian/rsc/RStudentCompanion.html.

Learning one basic approach to data reading and data writing in R will go a long way. Additional data management techniques in R are then easier to learn and can be added as needed. The essential steps in the basic approach you will learn in this chapter are as follows:

1. Set up the data in the form of a text file in a folder of your choice in the directory of your computer. This is done using resources outside of R.

2. In R, designate a "working directory" for the R session. The working directory is the folder in your computer where R is to find the data and write output files, if any.

3. Use one of R's data-reading commands such as read.table() to read the file and bring the data into the R workspace in the form of a data frame.

4. "Attach" the data to the work session using the attach() command, making the columns of data available to analyze as vectors.

5. Massage, crunch, fold, spindle, mutilate the data; plot the results. If there are any analysis results you would like to print or save in a file, gather them together into a data frame using the data.frame() command.

6. Print the results on the console or printer and/or store the results as a file on your computer.

7. If you intend to continue the R session using new data, "detach" the previous data using the detach() command.

Let us go through each of these steps in detail, using the college GPA data as an example.

1. Setting up a text file of data. Before starting R, download and take a look at the text file containing the college GPA data, using whatever simple text editor you normally use on your computer operating system (perhaps Notepad in Windows). We use a plain text editor instead of a sophisticated word processor to avoid having the word processor insert any extraneous formatting characters. The file looks like what is printed in Table 5.1.

The first line of the file is a line of text listing the names of the variables. The next lines are the data, in columns, separated by one or more spaces.

The file is a "space-separated text file." It is the format I prefer for my own work because entering data into a file in this format is easy with a simple text editor, and the format is highly portable to many analytical software programs.

Other common formats for data files are comma-separated text files (usually having a filename extension of .csv), tab-separated text files, and the widespread Microsoft Excel files (.xls or .xlsx). Space-, comma-, and tab-separated text files are all easy to handle in R, but Excel files need some preparation.

The R user community has contributed routines to read commercial file formats (like Excel, Minitab, and SAS) directly into R. However, in most circumstances, I recommend converting commercially formatted data into a space-separated (or comma- or tab-) text file before reading the data into R. That way, the R scripts you use will be simple and easier to follow, and you can reuse just a small handful of scripts with minor modifications to do all your data management. Most of the commercial packages contain some way to save data in some sort of plain text formatted file.

A data set in Excel, for instance, can be saved as a text file with the following actions. In Excel, from the "office button," select Save As→Other Formats to bring up a file-saving window. In the file-saving window, the Save as type menu has a variety of file formats to choose from. In the menu, space-separated text files are called Formatted text (Space delimited) (*.prn), while tab-separated text files are Text (Tab delimited) (*.txt) and comma-separated text files are CSV (Comma delimited) (*.csv). For the data file, choose a format, a filename, and a folder in your directory where you want the data stored. Excel automatically gives the extension .prn to space-separated files. I usually prefer the extension .txt because to me space-separated represents a text file, and so after creating the file with Excel, I usually find the file using Windows Explorer and change the extension. The extension and file name will not matter to R as long as they are designated properly in the R script for reading the file.

If you are typing data or downloading data from a web site, use a plain text editor to arrange the data into a form similar to the college GPA data (Table 5.1). Occasionally, data from the web or other sources contain special formatting or extended font characters, as is the case for Microsoft Word files. You can strip the characters by opening the file in Microsoft Word, highlighting (selecting) and copying the data to the computer clipboard, and then pasting the data into the plain text editor.

If you start tackling large data entry projects, having a text editor that allows copying and pasting of columns and rectangular regions of data is quite handy. A free one is Notepad++, and minimal searching will turn up other free or low-priced shareware text editors with this and other features. You might even find that you prefer to use an outside text editor for writing your R scripts, running them by simply pasting the scripts or script portions into the R console or by using the source() command (Chapter 2). (Start up a conversation with any geeky-type person by asking what text editor they prefer for programming! Do not ask about fonts until at least the third date, however.)

We assume from here on that the data you want to analyze in R exist as a text file in a folder somewhere in your computer. Get this task done for the college GPA data from Table 5.1. In the following examples, we will suppose that the file name for the college GPA data is GPAdata.txt.

2. Designate a working directory for R. R needs to know where to look for files to bring in and where to store files as output. There are two main ways to specify the working folder in R.

The first way to specify the working folder is the script method. You put a setwd() (set working directory) command in a script before any data-reading or data-writing commands occur. As an example, if the Windows directory path to the folder is c:\myname\myRfolder, then the syntax for setwd() is as follows:

```
setwd("c:/myname/myRfolder")
```

Notice the forward-leaning slashes. The world of scientists tends to be a Unix world. If you are working in Windows, R will find your folder just fine, after you pay proper tribute in your R statements with forward slashes.

If folder names have blank spaces in them, each blank space is indicated with a backward slash, followed by the space. The backward slash has the particular purpose in R (and Unix) of indicating that the character following it has a special meaning, in this case, that the space does not signify the end of the folder name. Suppose the folder is c:\My R Programs\Chemistry homework. The setwd() command should be as follows:

```
setwd("c:/My\ R\ Programs/Chemistry\ homework")
```

The second way to specify the working folder is the console method. You do this simply with the pull-down menu in the console: clicking File→Change dir... brings up a file directory window where you can select or change the working folder for your R session. Alternatively, you can just issue the setwd() command with the file path as the argument, as above, at the console prompt to accomplish the same thing mouse-free. With the console method, you do not need a setwd() command in your scripts.

Either way of designating a working folder in R has advantages. The first way specifies the working folder inside a script and is good for when you might want the script to be more self-contained or stand-alone, such as when you intend it as something to share with others. The second way specifies the working folder outside of any scripts (and before any scripts that need the folder are run) and is good for when you might be running several scripts to do work in the same folder.

3. Use one of R's data-reading commands. A main command for reading data from a text file into R is the read.table() command. The command works specifically with space-, tab-, or comma-delimited files. Set your working folder to be where GPAdata.txt is located, using the above console method. Then, try the read.table() command at the console:

```
> GPA.data=read.table("GPAdata.txt",header=TRUE,sep=" ")
> GPA.data
```

The contents of the file called GPAdata.txt are echoed at the console. They have become a data frame in your R workspace named GPA.data.

The header=TRUE option in the read.table() statement tells R that the first line in the data file is a header with text names corresponding to the columns of the data. The text names will become vector names in R after the data frame is attached (see step 4).

If the data do not have a header line, use header=FALSE. For instance, if the GPAdata.txt file had no header, you could add the names of the columns with an additional argument in the read.table() statement of the form col.names=c("univGPA", "ACT", "hsGPA", "sex", "housing"). Without the col.names= option, the columns of the data frame will automatically be called V1, V2, and so on (V for variable).

The sep=" " option in the read.table() statement tells R that a space or tab character separates data entries in each line of the input file. One could put a different text character between the quotes, such as a comma, if the file is comma-separated.

4. Attach the data frame. Although the data frame is in the workspace, no vectors have been defined yet. For instance, the header of the university GPA in the data file is univGPA. Type that at the console:

```
> univGPA
Error: object 'univGPA' not found
```

R reports that the vector named univGPA does not exist. Now, at the console, issue the following commands:

```
> attach(GPA.data)
> univGPA
```

You should see all the numbers in the univGPA column of the data set printed in the console. The columns of data (variables) have become vectors in R. The data are ready for analysis!

5. Analyze the data. Put results for printing or storing in a data frame. Graphical displays will often be the objective of many analyses performed.

We have already seen how to save graphs as files in various graphical formats. Sometimes, R analyses will produce additional numerical or categorical data, usually in the form of new vectors in the workspace. A good way to save that information is to collect the vectors together into a data frame and then store the data frame on the computer system as a text file.

I suspect you are beside yourself with curiosity at this point. Let us look right away at the relationship between college GPA and ACT scores. Two quantitative variables: sounds like a job for a scatterplot. Find out the answer at the console prompt:

```
> plot(ACT,univGPA,type="p",xlab="composite ACT",ylab="university
  GPA")
```

Revealed in the resulting graph is the dirty little secret of college admissions (Figure 5.1). Actually, it is a secret only to the test-takers and the general public. The testing companies know it (as documented in their own online reports), the college admission offices know it (from their internal studies), and the field of social science research knows it (hundreds of scientific journal articles). The data in support of the secret are extensive, going far beyond a few students in a statistics class.

The "secret" is that scores on the ACT or SAT are poor predictors of college success.

For the college GPA data, drawing some additional scatter plots and side-by-side boxplots for other variables in the data is the topic of a computational challenge at the end of the chapter.

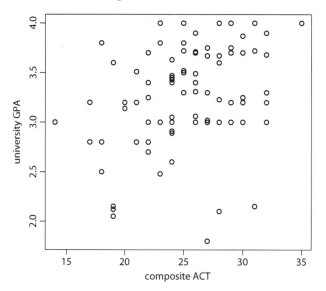

FIGURE 5.1
Scatterplot of composite ACT scores (horizontal axis) and university cumulative GPAs (vertical axis) for 90 undergraduate university students in an introductory statistics class.

For now, let us turn to the process of writing data from R to a text file. Suppose for illustration we set ourselves the simple task of separating the data into males and females. The male data will be stored in a file, and the female data will be stored in another file. Along the way, we can draw scatter plots of the college GPA and ACT variables for males-only and females-only to see if prediction might be improved by taking sex into account.

You will remember from Chapter 3 how to use logical vectors to pick out particular elements from data vectors. Let us pick out the elements from each vector corresponding to males and store them as separate vectors:

```
> univGPA.male=univGPA[sex=="m"]
> ACT.male=ACT[sex=="m"]
> hsGPA.male=hsGPA[sex=="m"]
> housing.male=housing[sex=="m"]
```

For instance, in the first statement, the sex=="m" statement produces a logical vector with TRUE corresponding to male and FALSE corresponding to female. The expression univGPA[sex=="m"] produces the vector of elements picked from univGPA that correspond to males only. The resulting vector is assigned the name univGPA.male.

You may plot ACT.male and univGPA.male now to see if the prediction is better for males only. If you find an improved prediction, the testing companies might want to hire you. If not, they might pay you to remain quiet.

Before storing the male-only data in a text file, we need to produce a new data frame consisting of the male-only data. The data.frame() command is the answer. Use it to "gather" a collection of vectors into a data frame. Issue the following commands at the console:

```
> GPA.data.male=data.frame(univGPA.male,ACT.male,hsGPA.male,
  housing.male)
> GPA.data.male
```

The data.frame() command combined the vectors designated as arguments within the parentheses into a data frame. The new data frame is named GPA.data.male.

We are now ready to store the data as a text file for future use.

6. Store the results from R analyses in a text file. If you want to print the new data frame, just type its name at the console:

```
> GPA.data.male
```

R will display the contents of the data frame at the console, formatted for nice printing. Highlight the contents on the console, and in the File menu, click Print. Alternatively, copy the highlighted contents into a separate text editor for printing.

The file-writing counterpart to the read.table() command is, appropriately, the write.table() command.

```
> write.table(GPA.data.male,file="GPAdata_male.txt",sep=" ")
```

The above write.table() command takes the data frame named GPA .data.male and saves it as a file named GPAdata _ male.txt in the current working directory of R. The sep=" " option in the read.table() statement tells R to use a space character to separate data entries in each line of the output file. One could put a different text character between the quotes, such as a comma, if desired. The names of the variables will be used as column names in the output data file.

The naming convention I like is to use periods in the names of R objects and underscores in the names of computer system files (except for the extension) and folders. You can invent and use your own conventions!

Without ending your R session, look outside of R into your working R folder to see that a new file named GPAdata _ male.txt has been placed there. Open the file with a text editor to check what is inside. If the new file seems as expected, close it, go back to the R session, and continue on. If not, um, well, close it, go back to the R session, and try to puzzle out what went wrong.

A computational challenge at the end of the chapter is to produce a similar stored text file containing the female-only data.

7. Clean up your mess. If further analyses with different data are contemplated for this R session, you might consider detaching any data frames from the workspace. That will help avoid errors arising from getting vector names confused. The detach() command is the tool:

```
> detach(GPA.data)
```

Of course, you can always just turn off R and turn it back on again to get a fresh clean workspace.

Final Remarks

What if there are 50 variables in a data frame and we want to pick out a subset such as males-only—do we have to type a statement with [sex=="m"] for every variable? That would work, but no. There are shorter and more elegant ways in R for picking out a subset of data from a data frame. They are displayed in Chapter 7. Typing, you are gathering by now, is a major anathema for R users.

WHAT WE LEARNED

Data sets in R are treated as special objects called data frames. A data frame is basically a collection of numerical or categorical variables (vectors). A text file of data can be brought into R and set up as a data frame for analysis. Analysis results can be collected into a data frame and then written as a text file in the computer system or printed.

Let us summarize what we learned so far about data input and output in the form of an R script. You can save the script as an example or template for use in future data-processing tasks:

```
#============================================================
# R script to read text file containing college GPA data
# into a data frame, extract the males-only data, and
# store the males-only data in a separate text file.  The
# input file should have data elements separated on lines
# by spaces, tabs, or commas.
#============================================================

#------------------------------------------------------------
# 1.  Designate the working folder where R is to find
# and write the data.  User must substitute the appropriate
# directory and file name between the quotes.
#------------------------------------------------------------
setwd("c:/myname/myRfolder")  # Precede any spaces in
                              # folder or file names with
                              # backslash, for example:
                              # setwd("c:/myname/my
                              # \ R\ folder")
#------------------------------------------------------------
# 2.  Read the data in the text file into an R data frame
# named GPA.data.  The header=TRUE statement tells R that
# the first line of the text file contains column or variable
# names.  The sep=" " option tells R that spaces separate
# data elements in lines of the text file.
#------------------------------------------------------------
GPA.data=read.table("GPAdata.txt",header=TRUE,sep=" ")
                              # Use header=FALSE if
                              # the first line of the
                              # data file does not
                              # contain column names.
#------------------------------------------------------------
# 3.  Attach the data frame, making the columns available in R
# as vectors.
#------------------------------------------------------------
attach(GPA.data)
```

```
#------------------------------------------------------------
# 4.  Extract the males-only data from the vectors.  (Also
# perform on the data whatever analyses are desired.)
#------------------------------------------------------------
univGPA.male=univGPA[sex=="m"]
ACT.male=ACT[sex=="m"]
hsGPA.male=hsGPA[sex=="m"]
housing.male=housing[sex=="m"]


#------------------------------------------------------------
# 5.  Gather the males-only vectors into a separate data
# frame.
#------------------------------------------------------------
GPA.data.male=data.frame(univGPA.male,ACT.male,hsGPA.male,
  housing.male)


#------------------------------------------------------------
# 6.  Store  the males-only data in a text file in the current
# working folder.  Data values in each line will be separated
# by whatever text character designated in the sep=" " argument.
# Print the males-only data in the console just by typing the
# name of the data frame.
#------------------------------------------------------------
write.table(GPA.data.male,file="GPAdata_male.txt",sep=" ")
GPA.data.male


#------------------------------------------------------------
# 7.  Detach the data frame, removing the vectors from
# workspace, if further analyses with different data are to be
# performed in the R session.
#------------------------------------------------------------
detach(GPA.data)
```

Computational Challenges

5.1. Graphically explore the college GPA data (Table 5.1) for potential associations of univGPA with other variables besides ACT. The graphical methods from Chapter 3 are your tools.

5.2. Separate out the males-only observations in the college GPA data (Table 5.1). Graphically explore the males-only data for potential associations of univGPA with other variables besides ACT.

5.3. Separate out the females-only observations in the college GPA data (Table 5.1). Graphically explore the females-only data for potential associations of univGPA with ACT as well as other variables.

5.4. Separate out the females-only observations in the college GPA data (Table 5.1). Save the females-only data in a separate text file on your computer system.

5.5. Enter the college GPA data (Table 5.1) into a Microsoft Excel spreadsheet (or other spreadsheet software). This can be done by copying and pasting or by opening the data text file from within Excel. Save the spreadsheet as an Excel file and close the spreadsheet. Now, open the resulting Excel file from within Excel and save it on your computer system as a space-separated text file (with a name different from the original). Compare the original text file and the Excel-filtered text file. Carefully document the steps for getting data from R into and out of Excel in a bulleted paragraph and store the paragraph as a document on your computer system for future reference. Share the document with coworkers.

Afternotes

Incidentally, if you read in a scientific journal article about data that interest you, most scientists will honor a polite request from you for the data (which typically appear in the article in graphs rather than raw numbers). There are some exceptions. First, investigators who conduct an important empirical study are usually granted a couple of years by the science world to study their data and present their own analyses of the data. Second, if the data requested are a subset of a big database, the extraction of the particular data that you want might be too onerous a programming project to fit into a busy scientist's schedule. Third, if you are perceived to be trolling to disrupt or hinder the scientist's work, your request will be received coldly.

Even so, most scientists pride themselves on openness in the practice of science. Many scientific fields and journals have various sorts of online data depositories (but you might need access to a research library network to obtain data files). Much of the data collected in research projects funded by the U.S. Federal Government are supposed to be openly available. And, most empirical scientists who might spend countless years in painstaking data collection will be more than happy to play Tycho to your Kepler, provided your seminal publication cites the data source prominently.

6

Loops

Do you see a pattern in the following sequence of numbers?

$$1 \quad 1 \quad 2 \quad 3 \quad 5 \quad 8 \quad 13 \quad 21 \quad \ldots$$

Of course you do; the sequence is the famous Fibonacci sequence that played a big role in Dan Brown's best-selling novel *The Da Vinci Code* (2003). We will use the sequence as an example for learning about loops in R.

First, we will need to know another way to pick out individual elements of a vector. We saw in Chapter 4 the method of picking out individual elements using logical (TRUE FALSE) vectors. Here, we will learn the index number method. In the R console, enter the first few numbers of the Fibonacci sequence into a vector:

```
>Fib=c(1,1,2,3,5,8,13)
>Fib
>[1] 1 1 2 3 5 8 13
```

Now, type the following command at the console, being careful to use square brackets:

```
>Fib[3]
>[1] 2
```

The third element in the vector Fib is 2. The seven elements in Fib can be referred to by their respective index numbers 1, 2, 3, 4, 5, 6, 7. Now, try this:

```
>Fib[c(3,4,6)]
>[1] 2 3 8
```

Do you see what happened? Here, c(3,4,6) is a vector with the numbers 3 4 6; when that vector is used as indexes inside the square brackets belonging to Fib, the third, fourth, and sixth elements of Fib are picked out.
Try:

```
>Fib[2:5]
>[1] 1 2 3 5
```

Remember that 2:5 produces the vector with the elements 2 3 4 5. Using it inside the square brackets picked out the second through fifth elements of Fib.

In general, if x is a vector and i is a positive integer, then x[i] is the *i*th element in x. Moreover, if x is a vector and y is a vector of positive integers, then x[y] picks out elements of x (those designated by the integers in y) and forms them into a new vector.

Writing a "For-Loop"

Let us set a goal of calculating the first, say, 50 members of the Fibonacci sequence. You can change this later to 1000 just for fun.

The keys to such calculation are (1) to understand the Fibonacci sequence as a mathematical recursion and (2) to use a loop structure in R to perform the calculations. Suppose we represent the Fibonacci numbers in the sequence as the symbols $r_1, r_2, r_3, \ldots$. So, $r_1 = 1$, $r_2 = 1$, $r_3 = 2$, and so on. Mathematically, each new Fibonacci number is related to the two preceding Fibonaccci numbers. We can write a recursion relation for the Fibonacci numbers as follows:

$$r_{i+1} = r_i + r_{i-1}.$$

Here, r_{i+1} is the "new" number, r_i is the "present" number, and r_{i-1} is the "past" number. To start the recursion rolling, we have to set the values of the initial two numbers: $r_1 = 1$, $r_2 = 1$.

One of the loop structures available in R is called a **for-loop**. A for-loop will repeat a designated section of an R script over and over.

Our computational idea is to set up a vector to receive the Fibonacci numbers as they are calculated one by one. An index number will pick the current and the past Fibonacci numbers out of the vector so that the new Fibonacci number can be formed by addition according to the recursion relation.

Start a new script in R. In the R editor, type the following commands:

```
num.fibs=50
r=numeric(num.fibs)
r[1]=1
r[2]=1
for (i in 2:(num.fibs-1)) {
    r[i+1]=r[i]+r[i-1]
}
r
```

Before you run the script, let us examine what happens in the script line by line. The first line defines a quantity named num.fibs, which will be the number of Fibonacci numbers produced. The second line defines r as a vector with length equal to num.fibs. The numeric() function used in the second line builds a vector of length given by num.fibs in which every element is a 0. Here, we use it to set up a vector whose elements are subsequently going to be changed into Fibonacci numbers by the script commands.

The third and the fourth lines of the script redefine the first two elements of r to be 1s. These are the first two values of the Fibonacci series. The user of the script can play with these initial values and thereby define a different recursion. Such new initial values would generate a different sequence instead of the Fibonacci sequence.

The for-loop takes place in the fifth, sixth, and seventh lines. The fifth line announces the for-loop. The opening curly brace in the fifth line and the closing curly brace in the seventh line bracket any R statements that will be repeated over and over.

In the fifth line, an index named i is defined. Each time through the loop, i will take a value from the vector defined in the fifth line to contain the elements 2,3,4,..., num.fibs-1. The first time through the loop, i will have the value 2, the next time i will have the value 3, and so on. The quantity i can be used in the loop calculations.

The sixth line provides the R statement for the Fibonacci recursion. The statement calculates the next Fibonacci number from r[i-1] and r[i] and stores it as r[i+1]. For other calculations, we could have more than one R statement here in between the curly braces. This sixth line, and any other lines of statements between the curly braces, will be repeated over and over, as many times as are provided in the for-loop statement (line 5 in this script). We used the programmer's typographical convention of indenting any statements in the loop, similar to the convention for function definitions.

The closing curly brace in line 7 defines the end of the for-loop. Note: a frequent typo that will cause an error is to have one of the curly braces facing the wrong way.

Checking the Loop

Getting a loop to work right takes some care and patience. Programmers learn to love this work. One effective trick is to calculate some of the R statements manually for the beginning and for the end of the loop. Here is how we would do that for our Fibonacci loop. Going into the loop, the values of r[1] and r[2] would both be 1. The first time through the loop, the value of i is 2. Thus, the statement r[i+1]=r[i]+r[i-1] becomes r[3]=r[2]+r[1]. Because r[2] is 1 and r[1] is 1, the statement will assign the value 2 to r[3]. The next time through the loop, the value of i is 3. The statement r[i+1]=r[i]+r[i-1] becomes r[4]=r[3]+r[2]. Because r[3] is 2 and r[2] is 1, the statement will assign the value 3 to r[4].

Each time through the loop, a new Fibonacci number is created from the previous two and stored in the vector named r. Manually reviewing the last time through the loop helps to ensure that the loop will not crash at the end. We are filling up a vector one element at a time. We must have the index i end at the right value: if i ends too low, we will have some leftover elements in r remaining as 0s instead of Fibonacci numbers, and if i ends too high, we will exceed the number of elements in r, causing an error message.

The last time through the loop, i will have the value num.fibs−1. The recursion statement r[i+1]=r[i]+r[i−1] becomes r[num.fibs]=r[num.fibs−1]+r[num.fibs−2]. By ending the loop at i=num.fibs−1, we ensured that the last execution of the recursion statement would assign the last element in r a Fibonacci value.

If everything goes well when the script is run, the vector r will have 50 elements in it, each one a member of the Fibonacci sequence. All that remains is to print elements of r to the screen, which is accomplished by the eighth and last statement of the script.

Run the script. It did not work? You likely made an error typing somewhere. Find the bug and fix the script; you know the drill. And when it finally runs, you will have the sought-after knowledge of the first 50 Fibonacci numbers.

OK, Mr. Fibonacci…So What?

The Fibonacci sequence is an abstract mathematical curiosity and might not seem all that useful at first. But you can think of it as a sort of basic population growth model in which there are two kinds of individuals: young and old. Both types of individuals in this population have on average one offspring per year, which becomes added to the young class. Young survive 1 year and become old, and old just die.

Next year, the young individuals would be the sum of the offspring from young and old individuals from this year:

$$\text{young}_{i+1} = \text{young}_i + \text{old}_i.$$

Also, the old individuals next year would be the young from this year:

$$\text{old}_{i+1} = \text{young}_i.$$

The preceding statement means the same thing as saying that the old individuals this year are the young individuals from last year:

$$\text{old}_i = \text{young}_{i-1}.$$

Substitute young$_{i-1}$ for old$_i$ in the first population equation to get:

$$young_{i+1} = young_i + young_{i-1}.$$

Compare this expression with the Fibonacci recursion.

Real-World Example

The Fibonacci sequence might seem to be a rather simple and unrealistic population growth model. However, the basic idea of the recursion is easily altered to describe real wildlife populations. In fact, we will see here how a recursion has been used to study the potential extinction of an endangered species.

A frequent life history seen in wildlife species is for the animals to go through three basic life stages: juvenile, subadult, and adult. The juveniles are the offspring that have been born within one time period (typically a year) of the present. The subadults are mostly nonreproductive and are usually between 1 and 2 years old. The adults are 2 or more years old and are the reproductive members of the population. A typical pattern is for the juveniles to have a fairly low 1-year survival probability, with the subadults having a relatively higher survival probability and the adults generally having the highest chance of living one more year.

Suppose we denote the number of juveniles in the population at time t to be J_t, the number of subadults to be S_t, and the number of adults to be A_t. We will use the population at time t, characterized by the three numbers J_t, S_t, A_t, to project what the population might be at time $t+1$.

Let us start with the easiest stage to model, the subadults. The subadults at time $t+1$ will be the juveniles at time t who survive for 1 year. If the fraction of juveniles that survive for 1 year is denoted by p_0, the number of juveniles at time t who survive for 1 year would be $p_0 J_t$. The model for the subadults is then

$$S_{t+1} = p_0 J_t.$$

Next, we look at the adults. The adults in the population at time $t+1$ come from two sources: subadults at time t who survive for 1 year, and adults at time t who survive for 1 year. We denote the fraction of subadults who survive for 1 year as p_1 and the fraction of adults who survive for 1 year as p_2. The number of adults at time $t+1$ becomes the sum of the two sources of adults:

$$A_{t+1} = p_1 S_t + p_2 A_t.$$

Finally, we need a model for the juveniles. The juveniles at time $t+1$ will be those born to the A_t adults in the coming year. The details of these sorts of wildlife projection models vary here depending on the biological details of when during the year juveniles are born and how they depend on their parents for survival. Many wildlife populations have a fixed and relatively short breeding season each year. We can adopt the convention that the numbers in each stage will be counted annually by the wildlife biologists just after such a breeding season, so that the juveniles are newly born. Then, the juveniles at time $t+1$ will be the number of adults at time t who survive the year, multiplied by the average number of newborn offspring per adult. We can write this as follows:

$$J_{t+1} = fA_t,$$

where f is the product of the fraction of adults surviving and the average number of newborn offspring produced by an adults during the breeding period. The constant f is commonly called the net fecundity by ecologists.

Let us look at the completed model by collecting the three projection equations together. The equations collectively take the "current" population sizes J_t, S_t, and A_t and calculate the population sizes 1 year in the future given by J_{t+1}, S_{t+1}, and A_{t+1}:

$$J_{t+1} = fA_t,$$

$$S_{t+1} = p_0 J_t,$$

$$A_{t+1} = p_1 S_t + p_2 A_t.$$

Mathematically, the equations are recursions for three simultaneous sequences of numbers: the numbers of juveniles, subadults, and adults through time. If we have numerical values for the survival and fecundity constants (f, p_0, p_1, p_2) and initial number of juveniles, subadults, and adults (J_0, S_0, A_0), then the equations can be used to predict the future numbers of each life stage in the population.

I hope you can see by now that calculating and graphing the future numbers is a ready-made task for R! In the calculation, we will use a for-loop to fill up three vectors (we will call them J.t, S.t, and A.t) representing the numbers in each life stage of a wildlife population through time.

The Northern Spotted Owl is a threatened species that lives and nests primarily in old-growth forests of the Pacific Northwest. The following survival and fecundity constants for the Northern Spotted Owl were extracted by Noon and Biles (1990) from data from various field studies: $p_0 = 0.11$, $p_1 = 0.71$, $p_2 = 0.94$, $f = 0.24$. These constants apply to females only, a common convention in studies of populations that have a constant ratio of males and females. Accordingly, the quantities J.t, S.t, and A.t in the model will refer to the number of females of each stage in the population.

Our script can take the basic structure of the Fibonacci script above. The tasks are as follows. (1) Set the total number of years for the projection. (2) Define the survival and fecundity constants. (3) Preallocate three vectors to receive the population numbers. (4) Set initial population sizes for juveniles, subadults, and adults and store them as the initial elements of the three vectors. We will use at first the values $J_0 = 1200$, $S_0 = 800$, and $A_0 = 2000$. (5) Construct a for-loop with an index that will represent time. (6) Put the three population projection equations inside the for-loop, using the current value of the index to pick out the appropriate elements of the vectors and project them 1 year forward. (7) After the loop is finished, print the vectors on the R console.

Stopping at number 7 above would constitute a failure of imagination. This is R, after all, meant to make laborious computational tasks easy so that we are free to *interpret*. Why are we calculating these vectors? Don't we want to *see* whether the Northern Spotted Owl will thrive or not? Of course! Once calculated, the vectors practically scream for a graph. Let us add: (8) Draw a graph showing the juveniles, subadults, and adults through time. However, we might want to produce and examine just the numbers first, in order to set the axes of the graph to display the entire range of the numbers clearly.

Producing such a graph has a few subtasks, which we will remember from Chapter 4: (8a) Build a vector for the horizontal axis containing time values. (8b) Plot the juveniles versus the time values with a plot() statement. We will use different line types to distinguish the different age groups. (8c) Add the subadults to the graph with a points() statement. (8d) Add the adults to the graph with another points() statement.

If you are inspired, stop reading here and write out an R script of your own for the Northern Spotted Owl. Then continue reading and compare your script with this one:

```
#=============================================================
# R script to calculate and plot age class sizes through
# time for an age-structured wildlife population, using
# projection equations.  Demographic rates for the
# Northern Spotted Owl are from Noon and Biles
# (Journal of Wildlife Management, 1990).
#=============================================================
num.times=20

p0=.11      # Age-specific survival and fecundity rates.
p1=.71      #
```

```
p2=.94        #
f=.24         #


J.t=numeric(num.times)   # Vector for juveniles.
S.t=J.t                  # Vector for subadults.
A.t=J.t                  # Vector for adults.


J.t[1]=1200   # Initial age class sizes.
S.t[1]=800    #
A.t[1]=2000   #


for (i in 1:(num.times-1)) {
    J.t[i+1]=f*A.t[i]            # Recursion equations
    S.t[i+1]=p0*J.t[i]          #   for projection of
    A.t[i+1]=p1*S.t[i]+p2*A.t[i] #  age classes.
}
J.t       # Print the results to the console.
S.t       #
A.t       #


# Plot the age classes through time.
time.t=0:(num.times-1)
plot(time.t,J.t,type="l",lty=2,xlab="time in years",
  ylab="population size",ylim=c(0,2600))
points(time.t,S.t,type="l",lty=5)
points(time.t,A.t,type="l",lty=1)
```

In the plot() statement above, the *y*-axis limits were set to the interval
(0, 2600) with the ylim= option. I picked these limits by first running the
script statements preceding the plot() statement and printing J.t, S.t,
and A.t at the console to view the range of resulting population sizes. The
lty= option in the plot() and points() statements is the "line type"
option that sets the type of lines used in the graphs to be dashed (lty=2)
for juveniles, long dashed (lty=5) for subadults, and solid (lty=1) for
adults.

Run the Spotted Owl script. The resulting prediction of the fate of the
species is depicted in Figure 6.1. If the survival and the fecundity constants
are accurate and do not change over the time horizon of the projection,
the model predicts a long but inevitable decline of this population toward
extinction. The "do not change" assumption provides simultaneously a ray
of hope and a ray of despair: Wildlife biologists point out that not only are
the survival and the fecundity constants measured imprecisely, but also that
the constants also vary considerably from year to year and might undergo
long-term trends due to habitat and climate change. The population could
possibly increase. Or possibly not.

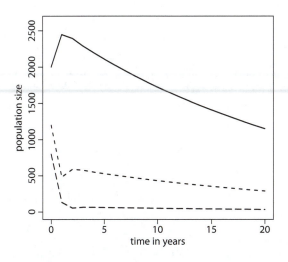

FIGURE 6.1
Female adult (solid line), female subadult (long dashes), and female juvenile abundances (short dashes) of the Northern Spotted Owl (*Strix occidentalis caurina*) projected for 20 years using survival and fecundity. (Data from Noon, B. R. and Biles, C. M., *Journal of Wildlife Management*, 54: 18–27, 1990.)

WHAT WE LEARNED

1. Elements of a vector can be picked out of the vector by using one or more indexes.

   Examples:

   ```
   > x=c(2,3,5,7,9,11,13,17,19)
   > x[8]
   [1] 17
   > x[1:5]
   [1]  2  3  5  7  9
   > y=2*x[c(1,3,5)]
   > y
   [1]  4 10 18
   ```

2. A "for-loop" repeats the execution of a designated section of a script over and over. The loop uses an index that takes a new value for each repeat. The index values come sequentially from a list of values provided as a vector in the "for" statement commencing the loop. The statements to be repeated are enclosed between curly braces.

   The example script sums the integers and the squared integers from 1 to k:

```
k=10
sum.int=1
sum.sqint=1
for (i in 2:k) {
    sum.int=sum.int+i
    sum.sqint=sum.sqint+i*i
}
sum.int              #  Print the sum of integers
k*(k+1)/2            #  Formula for comparison
sum.sqint            #  Print the sum of squared integers
k*(k+1)*(2*k+1)/6    #  Formula for comparison
```

Final Remarks

The for-loop in the "What We Learned" box to sum the integers and their squares is not really necessary in R to perform those particular calculations. I just used the loopy way to illustrate the syntax in R for writing the for-loop. Instead, in R, we can use vector thinking to accomplish the same thing more compactly and elegantly. Try the following script:

```
k=10
sum(1:k)
sum((1:k)^2)
```

In the more primitive nonvectorized computer languages, one must often go around and around in loops. With R and vector thinking, you can often avoid the rat treadmill.

Loops bring up the topic of speed of execution of an R script. The general guideline is that the same calculation takes longer with more lines of R statements. This guideline is especially important for for-loops. You can regard a for-loop as essentially many, many lines of statements. If the problem can be vectorized to avoid loops, then it will run faster in R. However, for problems of the sizes encountered in this book and in typical homework assignments in science courses, R is so fast that you will hardly notice a difference. If for some reason the course instructors become fully aware of the magnitude of the computing speed and the power that you command, then you can expect your assignments to grow larger and more sophisticated. (Shhh!)

Computational Challenges

6.1. a. If you assign new initial values in the Fibonacci script, you will get an entirely new sequence. Try computing the recursion for some different initial values. Name any interesting sequences you find after yourself and draw a graph for each one.

b. If you put some coefficients in the recursion, like

$$r_{i+1} = ar_i + br_{i-1},$$

where you choose the numbers a and b, then you will have an entirely new sequence. Alter the Fibonacci script to incorporate such coefficients and try computing the recursion for some different coefficient values and different initial values. Name any interesting sequences you find after yourself. If you want a sequence to have integer values, then you should use integers for the values of a and b. Nothing in the rules of Sequentiae prohibits negative values for a and b, however. If you choose real values for a and b, you will get sequences of real numbers.

6.2. Usher (1972) presented the survival and the fecundity values listed below for age classes of the blue whale. The blue whale lifespan was divided into seven age classes. Each time period, a fraction of the animals in each age class survive and enter the next age class. Animals in the last age class that survive each time period simply remain in the last age class. The blue whales start reproducing upon attaining the third age class, and so, each time period, several older age classes contribute newborns to the first age class.

Summarize the quantitative life history of the blue whale in a series of recursion equations, one equation for each stage. You will have to pick the symbols to represent the quantities in the equations. Then, plan and write an R script to project the population of blue whales for 20 time units. Draw one or more plots of the abundances of the age classes through time (if putting them all on one plot makes the graph appear too busy, draw several plots with two or three age classes on each plot). For the projection, use the following initial stage abundances for age classes 1–7 respectively: 200, 150, 120, 90, 80, 60, 100.

NOTE: A period of time in this model was defined to be 2 years, due to the slow growth of blue whales. In the equations and script, just allow time to increment ordinarily, like 0, 1, 2, 3, …, but realize that each time interval represents the passage of 2 years.

Age Class	1	2	3	4	5	6	7
Fraction Surviving	0.87	0.87	0.87	0.87	0.87	0.87	0.80
Fecundity	—	—	0.19	0.44	0.5	0.5	0.45

6.3. Plant ecologists often find it more useful to categorize a population of a plant species by stages based on size rather than age. The size-based stages often have a stronger relationship with mortality probabilities and fecundities. However, size stages have an extra feature in that some of the plants in a given stage will not grow enough in one time interval to enter the next stage. Each stage in the next time period will consist of some plants that were in that stage already and did not grow enough and some plants that were in the next smaller stage and grew enough to advance. Additionally, seeds for some species might not germinate during one time period and instead remain in the soil, viable for germination in some future time period.

Lloyd et al. (2005) collected data on the stages of black spruce (*Picea mariana*) at the northern end of the species range in Alaska. The stages they categorized were seed, germinant, seedling, sapling, small adult, medium adult, and large adult. The table below gives their data on (1) the fraction of individuals in each stage that remains in that stage during 1 year, (2) the fraction in each stage that moves on to the next stage in 1 year, and (3) the fecundities (seed production) of an average individual in each stage during 1 year.

Summarize the quantitative life history of black spruce in a series of recursion equations, one equation for each stage. You will have to pick the symbols to represent the quantities in the equations. Then, plan and write an R script to project a population of these plants for 20 time units. Draw one or more plots of the abundances of the size classes through time (if putting them all on one plot makes the graph appear too busy, draw several plots with two or three size classes on each plot). For the projection, use the following initial stage abundances: seeds 119627, germinants 7, seedlings 195, saplings 77, small adults 88, medium adults 69, and large adults 19.

Stage	Fraction Remaining in Stage	Fraction Advancing to Next Stage	Fecundities
seed	.493	.023	—
germinant	.029	.003	—
seedling	.918	.010	—
sapling	.964	.026	—
sm adult	.958	.032	479
md adult	.950	.040	479
lg adult	—	.991	479

References

Brown, D. 2003. *The Da Vinci Code*. New York: Doubleday.

Lloyd, A. H., A. E. Wilson, C. L. Fastie, and R. M. Landis. 2005. Population dynamics of black spruce and white spruce near the arctic tree line in the southern Brooks Range, Alaska. *Canadian Journal of Forest Research* 35:2073–2081.

Noon, B. R., and C. M. Biles. 1990. Mathematical demography of Spotted Owls in the Pacific Northwest. *Journal of Wildlife Management* 54:18–27.

Usher, M. B. 1972. Developments in the Leslie matrix model. In *Mathematical Models in Ecology*, ed. J. M. R. Jeffers, pp. 29–60. Oxford: Blackwell.

7

Logic and Control

We have seen that setting up scientific information in the form of one or more vectors is often convenient for graphing and analyzing the information, and that R is designed around the idea and use of vectors. R furthermore comes with a powerful set of tools for extracting and manipulating the information in vectors.

Logical Comparison Operators and Logical Vectors

We got a glimpse of **logical comparison operators** and **logical vectors** in Chapter 4, where we used a logical comparison operator to pick out Democratic and Republican unemployment rates out of a vector of unemployment rates. There, we used the "equal to" operator (==) to compare the elements in a categorical vector that contained "D" and "R" with a single text element (such as "D"). The result was a vector of logical elements TRUE and FALSE. That logical vector was in turn used to pick the desired elements out of the vector of unemployments. Here, we will take a more systematic look at logical comparison operators and logical vectors.

R has various logical comparison operators. Try the following three commands in the R console:

```
> x=c(3,0,-2,-5,7,2,-1)
> y=c(2,4,-1,0,5,1,-4)
> x<=y
[1] FALSE  TRUE  TRUE  TRUE FALSE FALSE FALSE
```

In the above, the first two statements just set up the vectors x and y. In the third statement, note especially the use of the "<=" symbol (an "is less than" sign typed next to an "equals" sign). That symbol in R is one of the logical comparison operators. It is the "less than or equal to" operator, and it compares every element in x with the corresponding element in y. If the element in x is indeed less than or equal to its counterpart in y, then the logical element TRUE is produced. Otherwise, when the element in x is greater than the element in y, a logical element FALSE is produced.

The result of the statement x<=y is a vector containing logical TRUE and FALSE elements giving the status of the "less than or equal to" relationship of each pair of x and y numbers. The vector is a logical vector, in which R is considered neither categorical nor numeric. The elements TRUE and FALSE are not entered with quotes around them, differing from how text elements are entered into categorical vectors. If you want, you can give the logical vector a name, say, compare.xy:

```
> compare.xy=(x<=y)
> compare.xy
[1] FALSE   TRUE   TRUE   TRUE FALSE FALSE FALSE
```

The parentheses in the first statement above are not really necessary; they just make the statement more readable.

Here is a list of the logical comparison operators available in R:

- < less than
- > greater than
- <= less than or equal to
- >= greater than or equal to
- == equal to
- != not equal to

The logical comparison operators all work in similar ways. As an example, look at the statement x>=y. If x is a numeric vector and y is scalar (a single number), the result is a logical vector with the same number of elements as x, in which each element of x has been compared to the number y:

```
> x=c(1,3,5,7,9)
> y=2
> x>=y
[1] FALSE   TRUE   TRUE   TRUE   TRUE
```

If y is a vector that has fewer elements than x, the values of y will be recycled in order until every element in x gets a comparison:

```
> x=c(1,3,5,7,9)
> y=c(2,4)
> x>=y
[1] FALSE FALSE   TRUE   TRUE   TRUE
```

In the above statements, 1 was compared with 2, 3 was compared with 4, 5 was compared with 2, 7 was compared with 4, and 9 was compared with 2. And, of course, if y is the same size as x, the result is a comparison of every element in x with the corresponding element in y.

You can compare categorical vectors with each other. "Greater than" and "less than" refer to the alphabetical order of the text strings, with "greater than" meaning closer to the letter z and "less than" meaning closer to the letter a. Try it:

```
> a=c("ann","gretchen","maria","ruth","wendy")
> b=c("bruce","ed","robert","seth","thomas")
> a>=b
[1] FALSE   TRUE FALSE FALSE   TRUE
> a<=b
[1]   TRUE FALSE   TRUE   TRUE FALSE
```

As with numerical vectors, the elements from a smaller categorical vector will be recycled to fill up all the remaining comparisons with a larger categorical vector.

Boolean Operations

The fun begins when you combine logical comparisons in Boolean operations. If you are adept at searching the Internet, you have no doubt become familiar with the Boolean operations "and," "or," and "not." For instance, you might want to search for web sites that have reviews of a new hit movie called *Carrot Cake Diaries* without getting extraneous web sites in the search list full of recipes or personal web logs. Similar problems occur when you want to pick out cases/subjects from a data frame having combinations of characteristics in common, such as the subclass of people in a survey who smoke as well as who are in favor of the death penalty.

The symbols for Boolean operations in R are

    &   and (ampersand character)

    | or (vertical bar character)

    !   not or negation (exclamation character)

The Boolean operators & and | connect two logical comparisons and return TRUE or FALSE depending on the joint truth or falsity of the two logical comparisons. The & returns TRUE if *both* logical comparisons are true, and | returns TRUE if *either* comparison is true. If the logical comparisons are vectors, then the Boolean operators will return a vector giving the outcomes for the corresponding pairs of comparisons. Here are some examples:

```
> x=c(3,0,-2,-5,7,2,-1)
> y=c(2,4,-1,0,5,1,-4)
> (x-y>-2) & (x-y<2)
[1]   TRUE FALSE   TRUE FALSE FALSE   TRUE FALSE
```

```
> x[(x-y>-2) & (x-y<2)]
[1]    3  -2   2
> (x-y>=2) | (x-y<=-2)
[1] FALSE  TRUE FALSE   TRUE   TRUE FALSE   TRUE
> x[(x-y>=2) | (x-y<=-2)]
[1]    0  -5   7  -1
```

In the above, (x-y>-2) & (x-y<2) produces a logical vector indicating which differences of x and y elements are greater than –2 as well as less than +2; another way of expressing this is that the elements of x and y are strictly within a distance of 2 from each other. The statement x[(x-y>-2) & (x-y<2)] picks out the elements of x, which are strictly within a distance of 2 from the corresponding elements of y. Also, (x-y>=2)|(x-y<=-2) produces a logical vector indicating which differences of x and y elements are *either* greater than or equal to +2 *or* less than or equal to –2; another way of expressing this is that the elements of x and y are at a distance of 2 or more from each other. And finally, x[(x-y>=2)|(x-y<=-2)] in the above statements picks out the element of x for which the distance to the corresponding y element is 2 or more.

The negation or not operator "!" (True. Not!) returns TRUE if the vectors compared do *not* obey the logical comparisons. Use of the negation operator basically reverses true and false:

```
> x=c(3,0,-2,-5,7,2,-1)
> y=c(2,4,-1,0,5,1,-4)
> (x-y>-2)
[1]   TRUE FALSE   TRUE FALSE   TRUE   TRUE   TRUE
> !(x-y>-2)
[1] FALSE   TRUE FALSE   TRUE FALSE FALSE FALSE
> (x-y>-2) & (x-y<2)
[1]   TRUE FALSE   TRUE FALSE FALSE   TRUE FALSE
> !((x-y>-2) & (x-y<2))
[1] FALSE   TRUE FALSE   TRUE   TRUE FALSE   TRUE
```

If the last statement was typed as !(x-y>-2) & (x-y<2), a different result would be returned because the ! operator would just negate the logical vector (x-y>-2) and not the whole statement. Liberal use of parentheses is recommended during sessions of heavy Boolean operating.

Stringing lots of Boolean operators together can get rather Boolean, rather fast. Care to work out by hand what will be the result of the following R statements? Try it, then see if you and R agree:

```
> x=c(3,0,-2,-5,7,2,-1)
> y=c(2,4,-1,0,5,1,-4)
> !(((x-y>-2) & (x-y<2)) | ((x-y<(-2)) & (x-y>2)))
```

In the last statement, the parentheses around the second –2 were necessary because R interprets the symbols "<-" as an assignment operation (like "=").

It is not that Boolean operators are not useful as well as not being easily interpretable, nor is it that Boolean operators that are not easily interpretable are not
useful; rather, it is that Boolean operators that are easily interpretable are useful.

Missing Data

Missing entries are not uncommon (apologies for the Boolean overdose) in
real data sets. There are many reasons data become missing, for example,
survey questions left blank, dropped test tubes, data recording errors, and so
on. Much of the time, graphical and statistical analyses used in science just
omit observations with missing data. Occasionally, sophisticated statistical
ways of estimating or simulating a missing observation can be employed. In
either case, the analysis software needs a way to recognize missing records
so that they can be processed in accord with the analyst's intentions.

The code for a missing data entry in R is NA. The code should be used in
vectors and data frames as a placeholder wherever a data entry is missing.
The vector calculations in R will return missing data values when the calculations are performed on vectors with missing values:

```
> u=c(3,5,6,NA,12,14)
> u
[1]   3   5   6 NA 12 14
> 2^u
[1]      8     32     64     NA   4096  16384
```

The prewritten analysis and graphing functions in R usually omit any
missing elements from the analyses:

```
> v=24*u+5
> plot(u,v)
```

Functions that do not just operate element-by-element but rather require
the entire vector for a calculation will come up entirely missing:

```
> mean(u)
[1] NA
> median(u)
[1] NA
> sqrt(u)
[1] 1.732051  2.236068  2.449490      NA  3.464102  3.741657
```

The missing data code NA is different from the "not a number" code NaN.
An NaN code results when some operation calculated for nonmissing elements is nonsensical, such as taking the square root of a negative number.

Another code, `Inf`, occurs when a calculation is either infinite, such as dividing by zero, or too big to handle in R's floating point arithmetic. The `Inf` code is slightly more informative than the `NaN` code, in that `Inf` indicates a directional infinite magnitude rather than just not making sense:

```
> 3/0
[1] Inf
> -3/0
[1] -Inf
> 10^(10^100)    # 10 to the googol power is a googolplex
[1] Inf
> (3/0)-(-3/0)
[1] Inf
> (3/0)+(-3/0)
[1] NaN
```

For the statement `(3/0)-(-3/0)`, there is some odd sense that `Inf` is a better result than `-Inf` or `NaN`, in that the distance between positive infinity and negative infinity can be thought of as positively infinite. By contrast, the statement `(3/0)+(-3/0)` represents positive infinity plus negative infinity, which has no sensible interpretation.

More about Indexes

We saw in Chapter 6 that indexes in square brackets can be used to pick out elements from a vector. Here are examples:

```
> x=c(3,7,5,-2,0,-8)
> x[3]
[1] 5
> x[1:3]
[1] 3 7 5
> x[c(2,4,5)]
[1]  7 -2  0
> x[1+3]
[1] -2
> i=4
> x[i]
[1] -2
```

There are lots more things you can do in R with indexes. For instance, if you use a negative number as an index, then the corresponding element in the vector is *excluded*:

```
> x[-3]
[1]  3  7 -2  0 -8
```

```
> x[-c(2,4,5)]
[1]   3   5 -8
```

Also, if you do a calculation with vectors, then you can extract elements on-the-spot with square brackets and indexes:

```
> y=c(2,7,-5,4,-1,6)
> (x+y)[1:3]
[1]   5 14   0
```

In the above, make sure the vector calculation is in parentheses or else R will apply the square brackets just to the vector y. The above construction works fine if a logical vector instead of an index vector is in the square brackets.

Indexes can be used to extract portions of data frames. Let us make a little data frame to see how such indexing works. We will combine x and y from above with a categorical vector into a data frame:

```
> z=c("jan","feb","mar","apr","may","jun")
> monthly.numbers=data.frame(x,y,z)
> monthly.numbers
    x   y   z
1   3   2 jan
2   7   7 feb
3   5  -5 mar
4  -2   4 apr
5   0  -1 may
6  -8   6 jun
```

Think of the data frame as a rectangular array of elements, with some columns numeric and some columns categorical. In the square brackets, each element is represented by two indexes separated by a comma; the first index always designates the *row*, and the second index designates the *column*:

```
> monthly.numbers[4,1]
[1]  -2
```

The indexes can be vectors so that entire portions of the data can be extracted:

```
> monthly.numbers[1:3,c(1,3)]
   x   z
1  3 jan
2  7 feb
3  5 mar
```

The above statement extracted rows 1, 2, and 3 and columns 1 and 3. To extract an entire row or column, omit the other subscript:

```
> monthly.numbers[5,]
   x  y   z
5  0 -1 may
```

```
> monthly.numbers[,2]
[1]   2   7 -5   4 -1   6
> monthly.numbers[,3]
[1] jan feb mar apr may jun
Levels: apr feb jan jun mar may
> monthly.numbers[1:3,]
  x  y   z
1 3  2 jan
2 7  7 feb
3 5 -5 mar
```

Note in the above that when you extract just a column, the result is considered a vector. Vectors in R cannot contain mixed types of data. If you extract one or more rows, or even a part of a row, R defines the result to be a data frame.

Conditional Statements

You can have R execute a statement or block of statements conditionally using an `if` command. The `if` command takes the following form:

```
if ( condition ) {
    statement 1a
    statement 1b
        ⋮
} else {
    statement 2a
    statement 2b
        ⋮
}
```

In the above, the *condition* is a logical expression that returns the value TRUE or FALSE. The statements 1a, 1b, and so on are performed by R only if the *condition* is TRUE. Statements 2a, 2b, and so on are performed by R only if the condition is FALSE.

Open the R editor and run the following little script to see the `if` statement in action:

```
x=3
if (x<=2) {
  y=5
  z=5
} else {
  y=6
  z=6
```

```
}
y
z
```

The output on the R console will look something like this:

```
> x=3
> if (x<=2) {
+     y=5
+     z=5
+ } else {
+     y=6
+     z=6
+ }
> y
[1] 6
> z
[1] 6
```

You can see from the continuation prompts "+" that R considers the if statement to be just one big statement that includes the whole bunch of conditional statements. In fact, you can omit the curly braces after if or else when there is just one conditional statement to be executed:

```
> x=3
> if (x<=2) y=5 else y=6
> y
[1] 6
```

Also, you can omit the else portion of the if statement (everything from else onward) when there are no conditional statements to be executed when the condition is FALSE:

```
> x=3
> y=6
> if (x>=2) y=5
> y
[1] 5
```

To illustrate the if statement in a more useful setting, let us build a basic probability event simulator. The simulator will generate a vector of desired length with just 0s and 1s in it, with each element representing the outcome of a random event.

There are many types of events for which the outcome is random, and there are only two possible outcomes. Some examples are (1) an atom of a radioisotope either decays in a unit of time or it does not decay during the time; (2) a baseball player either gets a hit in a defined at-bat (the usual definition of an at-bat excludes things like walks, being hit by a pitch, and sacrifices, in which other players share a substantial portion of credit or responsibility for the outcome), or the player does not get a hit during the at-bat; and (3) an adult bird either survives during the coming year, or the bird does not survive the year. What we will do is simulate many random outcomes, as if the event is being repeated over and over. In other words, we will simulate many atoms awaiting decay, or many at-bats of a baseball player, or many adult birds facing survival or death.

For each atom or player or bird, we will roll some computer dice and list the outcome as a 1 or a 0. A 1 will correspond to whatever outcome we are counting: an atom decays, a player gets a hit, or a bird survives. We will use some prespecified probability p as the probability of a 1, and $1 - p$ will then be the probability of a 0. Different processes can be expected to have different values of p, so we will want to be able to set different values of p in our simulations.

Writing a simple R function will do the trick. The arguments for the function will be n, the number of repetitions of the random event, and p, the probability of a 1 on any given repetition. The output of the function will be a vector of 0s and 1s that represent the random outcomes of each repetition. Such a vector will be handy for subsequently calculating properties of the random process; for instance, the sum of the vector's elements will be the number of 1s that occurred and the proportion of 1s in the vector will be the fraction of times that the outcome 1 occurred.

The "computer dice" we will use is the **uniform random number generator**. A uniform random number generator produces a randomly picked decimal number between 0 and 1. We will draw a uniform random number and compare it to p. If the value of the uniform random number is less than or equal to p, we will say that a 1 occurred on the current repetition of the random event; otherwise we will say that a 0 occurred. The word "if" in the previous sentence clues us that we will need an `if` statement in the function to compare the computer random number with the value of p to determine whether to assign a 1 or a 0 as the element in the vector of outcomes.

The uniform random number generator in R is the function `runif()`. It returns a whole vector of uniform random numbers. The argument to the function is the length of the vector desired. Try it a couple of times at the console:

```
> runif(10)
 [1] 0.48697717 0.52018777 0.47295558 0.73029481 0.14772913
 [6] 0.16835709 0.39762365 0.49683806 0.95916419 0.05179453
> runif(10)
 [1] 0.83867994 0.11604194 0.64532194 0.09253871 0.32728824
 [6] 0.89761517 0.92497671 0.64707698 0.27995645 0.05726646
```

Each time `runif()` is invoked, it gives new random numbers.

Enough talk. A function is worth a thousand words. In the R editor, type, save, and run the following script:

```
outcomes=function(n,p) {
    u=runif(n)        # generate a vector of uniform random numbers
                      #    of length n.
    x=numeric(n)      # x will eventually hold the random 0's
                      # and 1's.
    for (i in 1:n) {
       if (u[i]<=p) x[i]=1 else x[i]=0   # ith element of x is 1
                                         # with probability p.
    }
    return(x)
}
```

In the script, the vector u inside the function holds n uniform random numbers. The if statement is used to compare the *i*th element of u to p. When the *i*th element of u is less than or equal to p, the *i*th element of x is assigned to be 1. Otherwise, the *i*th element of x is assigned to be zero. Now, in the console, play with your new function for awhile:

```
> n=30
> p=.25
> outcomes(n,p)
 [1] 1 0 1 0 0 0 1 0 0 0 0 0 0 0 1 0 0 0 1 0 0 0 0 0 0 0 1 1 0 0
> outcomes(n,p)
 [1] 1 1 0 0 0 0 0 1 0 0 0 0 0 0 0 0 0 0 0 1 0 1 1 0 0 0 0 0 0 0
> p=.75
> outcomes(n,p)
 [1] 1 1 1 1 1 0 1 0 1 1 0 0 0 1 1 1 0 0 1 0 1 0 1 1 1 1 1 1 0 1
> outcomes(n,p)
 [1] 1 1 1 1 1 1 1 1 1 1 0 0 1 1 1 0 1 0 0 1 0 1 1 0 1 1 1 1 0 0
```

Notice how each time you invoke the outcomes function, a new set of random 0s and 1s is generated. The proportion of 1s that results is not necessarily the value of *p*; rather, each individual 0 or 1 represents a new roll of computer dice, with probability *p* that a 1 occurs and probability $1-p$ that a 0 occurs.

An average major league baseball (MLB) player (excluding pitchers) has a batting average of around .260. This means that during any given at-bat, the player has a probability of .26 of getting a hit. However, in any short sequence of at-bats, the player might experience many hits or few hits. Let us simulate many, many sequences of 30 at-bats, and see how much the player's apparent batting average (proportion of actual hits out of 30 at-bats) will jump around.

We will need to summon up several of our R skills. We can write a script in which we embed our `outcomes()` function in a loop, say, to create 100 sets of 30 at-bats. With those sets of at-bats, we can fill up a vector with 100 apparent batting averages. Then, we can graphically depict the variability of the batting averages with some suitable graphical display, maybe a histogram or a stem-and-leaf plot.

Ready? Type the following into a clean editor window:

```
outcomes=function(n,p) {
   u=runif(n)       # generate a vector of uniform random numbers
                    #   of length n
   x=numeric(n)     # x will eventually hold the random 0's and 1's
   for (i in 1:n) {
      if (u[i]<=p) x[i]=1 else x[i]=0  # ith element of x is 1 with
                               #    probability p
   }
   return(x)
}
n=30       # number of at-bats in a set
p=.26      # average MLB player's batting average
num.sets=100    # number of sets of n at-bats to simulate
bat.ave=numeric(num.sets)  # will contain the batting averages
for (i in 1:num.sets) {
   bat.ave[i]=sum(outcomes(n,p))/n  # number of hits in n at-bats
                            #    divided by n
}
hist(bat.ave)      # histogram of batting averages
stem(bat.ave)      # stem-and-leaf plot of batting averages
```

In the script, the `outcomes()` function definiton was repeated at the beginning so that the script will be entirely self-contained. The vector bat.ave contains 100 apparent batting averages; each one is the result of 30 at-bats. Now, every time this script is run, a different set of batting averages, and different pictures, will result. A typical appearance of the histogram is given in Figure 7.1. The stem-and-leaf plot corresponding to the histogram of Figure 7.1 was printed at the console:

```
The decimal point is 1 digit(s) to the left of the |
  0 | 7
  1 | 00003333333
  1 | 77777777
  2 | 000000000000000003333333333333333
  2 | 777777777777777
  3 | 0000000000000003333333333
  3 | 77777
  4 | 0
  4 |
  5 | 0
```

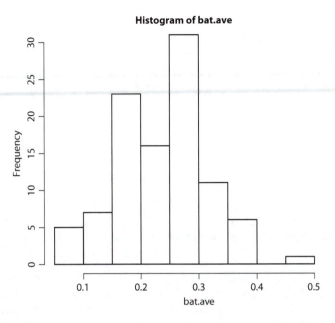

FIGURE 7.1

Histogram of 100 observed batting averages, each obtained from 30 simulated at-bats, from a hypothetical player with an underlying hitting probability of .260.

The striking feature of the plots is the variability. In quite a few stretches of 30 at-bats, an average player can be hitting over .300 (kind of a benchmark for star status) or even over .400 (ultra star status—.400 for an entire MLB season has not been accomplished since 1941). Also, in quite a few stretches of 30 at-bats, an average player can exhibit horrible runs of bad luck, hitting below .200 or even below .100. Think of an announcer commenting about how a player has been hitting really hot recently or has been in a terrible slump. Could it be that the player is the same as always and the hot streaks and slumps are just natural runs of luck?

Real-World Example

We will put our data management skills to the test in the following problem. The numbers in Table 7.1 are blood pressures, recorded in a clinical experiment to determine the best treatments for high blood pressure (Maxwell and Delaney 1990). Each number represents a subject in the study. The subjects were randomly assigned to different combinations of treatments, and the blood pressures were recorded at the end of the treatment period. The various treatments were in three different categories (called **factors** in experimental

TABLE 7.1

Blood Pressures of 72 Patients in a Clinical Experiment to Determine the Best
Treatments for High Blood Pressure.

Biofeed	Drug	Special Diet Yes						Special Diet No					
y	*a*	170	175	165	180	160	158	161	173	157	152	181	190
y	*b*	186	194	201	215	219	209	164	166	159	182	187	174
y	*c*	180	187	199	170	204	194	162	184	183	156	180	173
n	*a*	173	194	197	190	176	198	164	190	169	164	176	175
n	*b*	189	194	217	206	199	195	171	173	196	199	180	203
n	*c*	202	228	190	206	224	204	205	199	170	160	179	179

Note: Subjects were randomly assigned to different treatment combinations of drug (drug a,
            drug b, or drug c), special diet (yes or no), and biofeedback (yes or no).
Source: Maxwell, S. E., and H. D. Delaney, *Designing Experiments and Analyzing Data: A
            Model Comparison Perspective*, Wadsworth, Belmont, CA, 1990.

design): drug (drug a, drug b, or drug c), special diet (yes or no), and biofeedback (yes or no). Thus, there were 12 (= $3 \times 2 \times 2$) possible combinations of
treatments and 6 of the 72 subjects were assigned to each combination.

Let us calculate the mean blood pressures within each treatment combination and draw a comparative graph to help sort out which treatment
combinations were better than the others. The finished graph we will produce appears as Figure 7.2. The graph depicts the 12 mean blood pressures
as points on the vertical scale. The horizontal axis is not a numerical scale
but rather just identifies the drug type in the treatment combinations. The
point symbols identify the type of diet: circles are special diet "no," squares
are special diet "yes." The type of line connecting the points identifies the
biofeedback treatment: solid line is biofeedback "no" and dashed line is biofeedback "yes."

This type of graph, often called a **profile plot** in experimental design,
shows how the different factors interact, that is, how the strengths of the
effects of drugs a, b, and c on blood pressure depend on the other types of
treatments being administered. One can see at a glance in Figure 7.2 that
blood pressure under drugs a, b, and c can be low or high, depending on the
diet and biofeedback regimens, and the form of the dependence is inconsis-
tent from drug to drug. We will return to the graph and interpret its finer
points later; for now, we will concentrate on producing such a graph. It was
useful, though, to see the graph first to plan its production. When you are
launching into your own graphing projects, a hand sketch of the final graphical figure you want can help you plan the computational steps you would
need to include in an R script for drawing the figure.

We have a perplexing data management problem, though. The data file
corresponding to Table 7.1 is just a 6-by-12 array of numbers that does not
explicitly list the treatment combinations. Rather, the subjects' treatments are

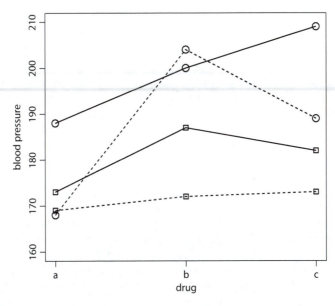

FIGURE 7.2

Plot of mean blood pressures for subjects under 12 different treatment combinations, from 6 subjects in each combination. Circles: special diet "no." Squares: special diet "yes." Solid line: biofeedback "no." Dashed line: biofeedback "yes."

recorded implicitly by the position of the blood pressure record in Table 7.1. As indicated in Table 7.1, the upper three lines of data correspond to biofeedback "yes" and the lower three lines are biofeedback "no." Also, the left six columns are special diet "no," and the right six columns are special diet "yes." Finally, the first three lines are, respectively, drugs "a," "b," and "c" and then the lines four through six are also, respectively, drugs "a," "b," and "c."

What we would like, for easier plotting, is to arrange these data into a data frame. The conventional arrangement of a data frame, in which each line corresponds to a subject and the columns correspond to variables, will help us issue simple and intuitive commands to analyze the data. We then will save the resulting table so that we might return to the data without having to puzzle out the data input problems again.

When we are done, the data frame should have 72 lines and look something like the following:

subject	bp	biofeed	diet	drug
1	170	y	n	a
2	175	y	n	a
3	165	y	n	a
⋮	⋮	⋮	⋮	⋮
72	179	n	y	c

Here, bp is a variable containing the subjects' blood pressures, and biofeed, diet, and drug are categorical variables indicating the type of treatment for that subject.

There are many different ways in R to construct the desired data frame. The usual advice for tackling such complex tasks is to break up the work into simple, well-understood steps, and worry about elegance later (like, next year). Our steps for building the data frame will be as follows: (1) Read the raw numbers into a preliminary data frame having 6 rows and 12 columns. (2) Stack the columns of the data frame into a vector named bp. (3) Use logical statements to build a text vector named biofeed containing "y" and "n" on the appropriate lines. Also, use logical statements to build a text vector named drug containing "a", "b", and "c" on the appropriate lines. Forming these two text vectors might possibly be combined in one step because they require similar logical tasks: they both are defined in terms of rows of the original table of raw numbers. (4) Use logical statements to build a text vector named diet containing "y" and "n" on the appropriate lines. This is likely a different logical task that of biofeed and drug because drug is defined in terms of columns of the original table. (5) Combine bp, biofeed, diet, and drug into a data frame. (6) Calculate the mean blood pressures within every treatment combination of biofeed, and drug, and diet. (7) Plot the data in the form of Figure 7.2. The plot itself will require a workout of our R graphing skills.

Here is a script to accomplish these tasks:

```
#=================================================================
# R script to read blood pressure data and draw a profile plot
# of the means under different treatment combinations.
#=================================================================

#-----------------------------------------------------------------
# 1.   Read the raw numbers into a preliminary data frame
# having 6 rows and 12 columns. The raw numbers are assumed
# to be in a space-delimited text file named
# "blood pressure data.txt" in the current working directory.
#-----------------------------------------------------------------
bp.data=read.table("blood pressure data.txt",header=FALSE)
#-----------------------------------------------------------------
# 2.   Stack the columns of the data frame into a vector named
# bp.
#-----------------------------------------------------------------
bp=c(bp.data[,1],bp.data[,2],bp.data[,3],bp.data[,4],
     bp.data[,5],bp.data[,6],bp.data[,7],bp.data[,8],
     bp.data[,9],bp.data[,10],bp.data[,11],bp.data[,12])

#-----------------------------------------------------------------
# 3.   Use logical statements to build a text vector named
```

```
# biofeed containing "y" and "n" on the appropriate lines.
# Also, use logical statements to build a text vector named
# drug containing "a", "b", and "c" on the appropriate lines.
#---------------------------------------------------------------
n.bp=length(bp)             # Number of observations in vector bp.
biofeed=character(n.bp)  # Vector of blank characters (" ").
drug=biofeed                # Vector of blank characters.
for (i in ((0:(n.bp/6-1))*6+1)) {
  biofeed[i:(i+2)]="y"              # Label biofeed y or n.
  biofeed[(i+3):(i+5)]="n"     #         --
  drug[c(i,i+3)]="a"              # Label drug a, b, or c.
  drug[c(i+1,i+4)]="b"          #         --
  drug[c(i+2,i+5)]="c"          #         --
}


#------------------------------------------------------------------
# 4.  Use logical statements to build a text vector named diet
# containing "y" and "n" on the appropriate lines.
#------------------------------------------------------------------
rnum=1:n.bp                   # Index vector from 1 to n.bp.
diet=character(n.bp)       # Character vector of length n.bp.
diet[rnum<=n.bp/2]="n"    # First half labeled n.
diet[rnum>n.bp/2]="y"     # Second half labeled y.

#------------------------------------------------------------------
# 5.  Combine bp, biofeed, diet, drug, into a data frame.
#------------------------------------------------------------------
bp.data.new=data.frame(bp,diet,biofeed,drug)

#------------------------------------------------------------------
# 6.  Calculate the mean blood pressures within every treatment
# combination of biofeed, drug, and diet.  Put the means in a new
# data frame named bp.means.  Give the variables names.
#------------------------------------------------------------------
bp.means=aggregate(bp,by=list(diet,biofeed,drug),FUN=mean)
                  # Applies mean() function to bp elements
                  # having identical levels of diet, biofeed,
                  # and drug.
names(bp.means)=c("diet.m","biofeed.m","drug.m","bp.m")

attach(bp.means)

#------------------------------------------------------------------
# 7.  Graph the means in a profile plot.
#------------------------------------------------------------------
plot(c(1,2,3),bp.m[(diet.m=="n")&(biofeed.m=="n")],type="o",
  lty=1,pch=1,cex=1.5,ylim=c(160,210),xlab="drug",ylab="blood
  pressure",xaxt="n")
Axis(at=c(1,2,3),side=1,labels=c("a","b","c"))
points(c(1,2,3),bp.m[(diet.m=="y")&(biofeed.m=="n")],type="o",
  lty=1,pch=0)
```

```
points(c(1,2,3),bp.m[(diet.m=="n")&(biofeed.m=="y")],type="o",
   lty=2,pch=1,cex=1.5)
points(c(1,2,3),bp.m[(diet.m=="y")&(biofeed.m=="y")],type="o",
   lty=2,pch=0)
```

There are some new techniques in the script. Let us examine the details.

Parts 1 and 2 are self-explanatory. In Part 3, categorical variables names biofeed and drug are coded. The value of n.bp is 72, which is the number of observations in the blood pressure data set. Now, envision the columns in Table 7.1 stacked into one big column, with the leftmost column on top. That stack is the variable bp. In this stack of blood pressures, the biofeedback and drug treatments repeat themselves naturally in groups of 6. Biofeedbacks for each group of 6 observations would be yyynnn, and the drug treatments would be abcabc. The vector ((0:(n.bp/6-1))*6+1) in the for-loop is the sequence 1, 7, 13, 21, …, 67 (you should check this). Thus, the index i increments by 6s. The first time through the loop, elements 1, 2, and 3 of biofeed are assigned the text character "y", while the elements 4, 5, and 6 are given the character "n". Also, elements 1 and 4 of drug get the character "a", elements 2 and 5 get the character "b", and elements 3 and 6 get the character "c". The loop repeats for the values i=7, 13, …, and each subsequent group of 6 elements of biofeed and drug gets assigned in similar fashion.

Part 4 builds the diet variable. This variable is easier because of the way the columns of the data were stacked. The first half of the elements of bp corresponds to diet="n", and the last half of bp corresponds to diet="y".

Part 5 of the script builds the data frame that we were after, with variables bp, diet, biofeed, and drug. At this point, one could optionally save this data frame as a file with a write.table() command, if further analyses were contemplated.

In Part 6, the mean blood pressures within each unique treatment combination are calculated, using a handy function called aggregate(). The aggregate() function splits the data into subsets, computes summary statistics for each, and returns the result in the convenient form of a data frame. The aggregate() statement in the script applies the mean() function to elements of bp that have the same combination of diet, biofeed, and drug levels. It takes as its arguments a variable (here bp), a list of categorical variables (or factors), and the name of the function to be applied. In the names= statement, the variables in the new data frame are given names diet.m, biofeed.m, and so on to distinguish them from the variable names in the full data.

Part 7 constructs the plot. In the plot() statement, the x-axis is taken to be simply the vector 1, 2, 3. The Boolean statement picks out the three elements of bp.m with no special diet and no biofeedback. These three elements have the three different drug treatments a, b, c. The plot() statement draws the three points with circles connected by a solid line. The option cex=1.5 draws the plotting characters (here, circles) 1.5 times larger than the default for better visibility. The option ylim=c(160,210) defines the y-axis to be between 160 and 210 to focus the graph on the range of the mean blood pressures (the values 160 and 210 were chosen by trial and error after running the script a few times). The option `xaxt="n"` suppresses the automatic drawing of tic marks on the *x*-axis. We are instead going to substitute custom tics labeled a, b, and c.

In the `Axis()` statement, the `at=` argument provides the *x*-axis with tic marks at the values 1, 2, 3. The `side=1` argument causes the tic marks to face outward from the plot region. The `labels=` argument provides the labels for the three tic marks.

Subsequent `points()` statements in Part 7 of the script add the other three unique combinations of `diet` and `biofeed` treatments. The statements provide different plotting characters and line types.

Run the script to reproduce Figure 7.2. Then, if you type bp.means at the console, you will see the contents of the data frame that was plotted in Figure 7.2:

```
> bp.means
   diet.m biofeed.m drug.m bp.m
1       n         n      a  188
2       y         n      a  173
3       n         y      a  168
4       y         y      a  169
5       n         n      b  200
6       y         n      b  187
7       n         y      b  204
8       y         y      b  172
9       n         n      c  209
10      y         n      c  182
11      n         y      c  189
12      y         y      c  173
```

Figure 7.2 shows that different combinations of blood pressure treatments in this experiment produced vastly different outcomes. The best combination of treatments (yielding the lowest mean blood pressure) was drug "a," biofeeback "yes" and special diet "no," with drug "a," biofeedback "yes," and special diet "yes" a very close second. The worst combination of treatments was drug "c," biofeedback "no," and special diet "no."

Final Remarks

The graph in Figure 7.2 practically screams for a legend to help the viewer keep track of the symbols. In Chapter 13, we will learn how to add legends, text, lines, titles, and other items to graphs.

WHAT WE LEARNED

1. The logical comparison operators for a pair of vectors x and y take the form x>y, x<y, x>=y, x<=y, x==y, and x!=y and are, respectively, greater than, less than, greater than or equal to, less than or equal to, equal to, and not equal to. Each operator compares corresponding elements of x and y and returns a logical vector (TRUE TRUE FALSE ...) based on the truth or falsity of the comparisons.

2. The Boolean operators & and | ("and" and "or") connect two logical comparisons and return TRUE or FALSE depending on the joint truth or falsity of the two logical comparisons. The & returns TRUE if *both* logical comparisons are true, and | returns TRUE if *either* comparison is true.

Example:

```
> x=c(4,5,6,7,8)
> y=c(1,2,3,4,5)
> z=c(2,3,4,5,9)
> (x>=y)&(x<=z)
[1] FALSE FALSE FALSE FALSE TRUE
> (x>=y)|(x<=z)
[1] TRUE TRUE TRUE TRUE TRUE
```

3. The Boolean operator ! ("not") returns TRUE if the vectors compared do *not* obey the logical comparisons.

Example:

```
> x=c(4,5,6,7,8)
> y=c(1,2,3,4,5)
> z=c(2,3,4,5,9)
> !((x>=y)&(x<=z))
[1] TRUE TRUE TRUE TRUE FALSE
```

4. The code for a missing data entry in R is NA. The code should be used in vectors and data frames as a placeholder wherever a data entry is missing. R uses the code NaN ("not a number") when a numerical operation is nonsensical or undefined. Additional codes Inf and -Inf are returned when a numerical operation is positively or negatively infinite. Examples:

```
>x=c(1,4,9,NA,-1,0,0)
> y=c(1,1,1,1,1,1,-1)
> y/sqrt(x)
[1] 1.0000000 0.5000000 0.3333333  NA  NaN  Inf -Inf
```

5. Elements can be selected from vectors or data frames with indexes in square brackets or logical vectors in square brackets. Calculations can be performed in the square brackets to produce the indexes or logical vectors. Negative indexes exclude rather than include the elements. Data frames use two indexes separated by a comma, with the first index indicating row number and the second indicating column number. Omit the row index to select all rows and omit the column index to select all the columns.

Examples:

```
> x=c(-3,-2,-1,0,1,2,3)
> y=c(-6,-4,-2,0,2,4,6)
> x[2:6]
[1] -2 -1  0  1  2
> x[c(1,3,5)]
[1] -3 -1  1
> x[-c(1,3,5)]
[1] -2  0  2  3
> x[x<=y]
[1] 0 1 2 3
> z=data.frame(x,y)
> z[2,2]
[1] -4
> z[c(2,3,4),2]
[1] -4 -2  0
> z[,1]
[1] -3 -2 -1  0  1  2  3
```

6. You can have R execute a statement or block of statements conditionally using an `if` command. The `if` command takes the following form:

```
if ( condition ) {
    statement 1a
    statement 1b
         ⋮
} else {
    statement 2a
    statement 2b
         ⋮
}
```

The statements in the `if` braces are executed only if the *condition* results in a logical TRUE. The `else {}` portion is optional and is executed if the condition is FALSE.

Examples:

```
> x=3
> if (x<=2) {
+    y=5
+    z=5
+ } else {
+    y=6
+    z=6
+ }
> y
[1] 6
> z
[1] 6
> if (x>=2) y=5 else y=6
> y
[1] 5
> y=6
> if (x>=2) y=5
> y
[1] 5
```

7. The function `runif(n)` is a uniform random number generator in R. It generates a vector of n random numbers between 0 and 1.

Examples:

```
> n=5
> runif(n)
[1] 0.50589617 0.25741635 0.03789863 0.85341966
[5] 0.57386399
> runif(n)
[1] 0.16857682 0.03718524 0.53016185 0.16660328
[5] 0.63506073
> x=c(1,2,3,4,5)
> x[runif(n)<=.5]    # Flips coin for each element of x.
[1] 2 3
> x[runif(n)<=.5]    # Again.
[1] 1 2 4
> y=c(-4,3,5,-2,7)
> rank(y)        # Ranks of elements of y
                 #  (1 is lowest, 2 is 2nd lowest, etc.).
[1] 1 3 4 2 5
> y[rank(runif(n))]  # Arrange elements of y in
  random order.
[1] -2  7 -4  5  3
> y[rank(runif(n))]  # Again.
[1]  7  3 -4 -2  5
```

In this chapter, the outcomes() function, which returns a sequence of random 1s or 0s, was not written as efficiently in R as it could have been. In Chapter 14 on probability models we will express such simulations in vector form, without the need for loops. R has many different probability simulators already built in vector form. But it is helpful at first to write things out in long form to understand the structure of the calculations better.

Computational Challenges

7.1. The following are some winning times (in seconds) of the Olympic men's 1500-m race through the years:

1900	1920	1936	1948	1960	1968	1972	1976
246.0	241.9	227.8	229.8	218.4	214.9	216.3	219.2

1980	1984	1988	1992	1996	2000	2004	2008
218.4	212.5	16.0	220.1	215.8	212.1	214.2	212.9

a. Put the numbers into the vectors year and time, and put the two vectors into a data frame named olympic.1500m.

b. Draw a scatterplot of the data, with year on the horizontal axis. In the scatterplot, identify the points from years 1900–1968 and 1972–2008 with different symbols.

c. Calculate the mean winning time for the years 1900–1968 and the mean winning time from 1972 to 2008.

7.2. For the GPA data set from Chapter 5, construct a profile plot of the mean university GPA (vertical axis) as it varies across different housing types (off-campus, on-campus, fraternity/sorority) on the horizontal axis. Show two profiles on the plot using two line types (maybe dashed and solid), representing the data separated by males and females.

7.3. Draw a profile plot similar to that in Question 7.2, except use mean ACT score on the vertical axis.

7.4. Draw a profile plot similar to that in Question 7.2, except use mean high school GPA on the vertical axis.

7.5. A basketball player has a long-run success probability of 60% for free throws. Simulate for this player 100 batches of 20 free throws. Draw a histogram of the results. How often is the player shooting really hot (say, 15 or more successes of 20)? How often is the player shooting really cold (say, 9 or less successes)?

Afternotes

The author has coached many youth baseball and fastpitch softball teams. A summer season of 10 games might involve as few as 30 at-bats per player. A perennial challenge has been to convince players, parents—and coaches— that 30 at-bats are nowhere near enough to determine who are the good hitters and who are not. With only 30 at-bats, a good hitter can easily display a terrible slump; a bad hitter can seem like the hottest player in the league. Players, and unfortunately coaches, form impressions of players' hitting abilities all too fast. Coaches and even players themselves will give up too quickly and assign, or resign themselves to, the fate of benchwarmer. The practice sadly extends to other youth sports as well.

Reference

Maxwell, S. E., and H. D. Delaney. 1990. *Designing Experiments and Analyzing Data: A Model Comparison Perspective*. Belmont, CA: Wadsworth.

8

Quadratic Functions

Since the time of ancient Greece, architects have admired a certain proportion for the height and width of a building (Figure 8.1). The proportion is based on a particular rectangle produced according to a particular rule. The rule states that if you cut the rectangle into a square and another smaller rectangle, the sides of the smaller rectangle have the same proportion as the original (Figure 8.2). The resulting rectangle is known as the **Golden Rectangle**, and the ratio of the larger side to the smaller is called the **Golden Ratio**.

Let us calculate the Golden Ratio. In Figure 8.2, the long length of the rectangle is taken to be x, and the square inside the rectangle will be defined to have sides of length 1 (which is the length of the rectangle's short side). Thus, the ratio of the long side to the short side, the Golden Ratio, is the quantity x. The quantity x will be a real number greater than 1.

The Golden Ratio x obeys a mathematical rule: the proportions of the newly formed small rectangle are the same as the proportions of the original rectangle. The long side of the original rectangle is x, and the short side is 1. The long side of the small rectangle is 1, and the short side is $x - 1$ (Figure 8.2). If the sides have the same proportions, that means the ratios of the long side to the short side are the same for both rectangles:

$$\frac{x}{1} = \frac{1}{x-1}.$$

This is an equation that we need to solve for x. On the left-hand side, $x/1$ is just x. We clear the denominator on the right-hand side by multiplying both sides by $x - 1$:

$$x(x - 1) = 1,$$

$$x^2 - x = 1.$$

We then get all the terms over to the same side of the equals sign by subtracting 1 from both sides:

$$x^2 - x - 1 = 0.$$

FIGURE 8.1
Parthenon, on the Acropolis in Athens, Greece, inscribed inside a Golden Rectangle.

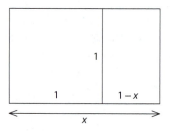

FIGURE 8.2
A Golden Rectangle can be cut into a square and another rectangle with sides having the same proportions as the original.

The remaining algebraic steps for solving for the value of x that satisfies this equation might not be apparent at a glance. Let us first draw a graph to look at the situation. The following script draws an x–y plot of the quantity $y = x^2 - x - 1$ for a range of values of x between xlo and xhi:

```
#  R script to draw the equation for the Golden Ratio.
xlo=-1
xhi=2
x=xlo+(xhi-xlo)*(0:100)/100
y=x^2-x-1
plot(x,y,type="l")
y2=numeric(length(x))
points(x,y2,type="l",lty="dashed")
```

With the script, I originally had to do some trial runs with different values of xlo and xhi to get the graph to look nice. You could try some other values. The statement x=xlo+(xhi-xlo)*(0:100)/100 calculates a vector x of values starting at xlo and ending at xhi (check this at the console if you do not

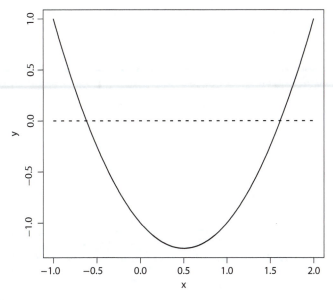

FIGURE 8.3
Solid curve: plot of the equation $y = x^2 - x - 1$. Dashed line: plot of $y = 0$.

see how this works). Then y is a vector of values of $x^2 - x - 1$ calculated for each value in the vector x. The plot() statement produces the curved solid line in Figure 8.3. The y2=numeric(length(x)) statement produces a vector y2 with the same number of elements as x that contains all 0's. Finally, the points statement draws the dashed line on the graph using the zeros in y2.

The resulting figure (Figure 8.3) shows a curve (solid line) representing $y = x^2 - x - 1$ and a dashed line drawn horizontally at $y = 0$. The two intersections of the solid curve and the dashed line occur where $x^2 - x - 1 = 0$. The intersections occur at two values of x that are potential solutions to our problem. However, one of the solutions is negative, around a little less than −.5. Obviously, the ratio of the rectangle sides cannot be negative, so we can discard that value for our purposes. The other solution, seen in Figure 8.3 to be a little greater than +1.5, is the physically meaningful solution to our Golden Ratio problem. The longer side of a Golden Rectangle is a little bit more than one and a half times the length of the shorter side.

Can we make the value of x more precise? Well, we can use our graph script to zoom in for a close-up, by simply running and rerunning the script, each time raising the value of xlo and lowering the value of xhi to catch the crossing point in between them, like a computer version of "hi-lo." After just a few rounds of this, I was able to squeeze the Golden Ratio value to five decimal places, using xlo=1.61800 and xhi=1.61805 (Figure 8.4). Incidentally, on the vertical axis in Figure 8.4 we encounter for the first time in this book numbers printed by the computer in the form of computer scientific notation.

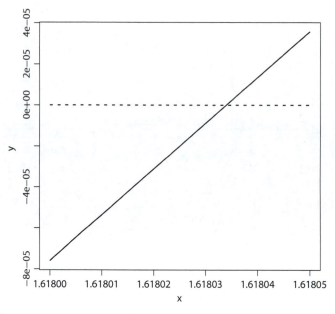

FIGURE 8.4
Solid curve: plot of the equation $y = x^2 - x - 1$. Dashed line: plot of $y = 0$.

In computer scientific notation, 3.4e03 means: $3.4*10^3$. Thus, 4e-05 at the top of the vertical axis is $4*10^{(-5)}$, which is the same as $4/10^5$. The numbers on the vertical axis of Figure 8.4 are pretty close to zero.

By continuing the hi-lo squeeze, we could get more decimal places than we could possibly need to design any building. But we can do even better, in the form of a mathematical formula.

The equation given by $y = x^2 - x - 1$ is a special case of a **quadratic function**. The general form of a quadratic function is given by

$$y = ax^2 + bx + c,$$

where a, b, and c are real-valued constants, with a not equal to zero. If a is zero, the function is a **linear** function. The term with x^2 (the "quadratic" term) gives the function its curvature. The Golden Ratio quadratic function has $a = 1$, $b = -1$, and $c = -1$.

The graph of a quadratic function is a curve called a **parabola**. If the value of a is negative, the parabola is "hump-shaped" and opens downward (like an upside down bowl), while if the value of a is positive, the parabola opens upward (like a rightside up bowl). Four parabolas appear in Figure 8.5. A computational challenge at the end of the chapter will invite you to draw these parabolas.

If the quadratic function intersects the horizontal axis (the dashed line in Figure 8.3), then there will be at most two intersection points. Those values of x

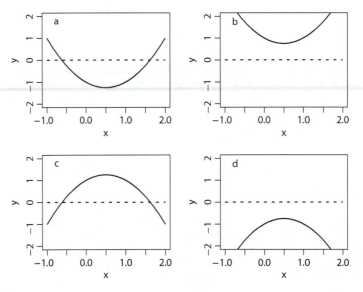

FIGURE 8.5
(a) Plot of the equation $y = x^2 - x - 1$ (solid curve). (b) Plot of the equation $y = x^2 - x + 1$ (solid curve). (c) Plot of the equation $y = -x^2 + x + 1$ (solid curve). (d) Plot of the equation $y = -x^2 + x - 1$ (solid curve).

at the intersection points are the roots of the quadratic function. For our Golden Ratio quadratic function, the larger of those roots is the Golden Ratio value.

The algebraic method of "completing the square" can be used to derive formulas for the roots of a quadratic equation (see the "Roots of a Quadratic Equation" box at the end of this chapter). The exercise is virtually compulsory in high school algebra and is reproduced in the box for a review. The results are those famous (notorious?) formulas, "minus bee plus or minus the square root of bee squared minus four ay cee over two ay," that have formed the no-more-math tipping point for more than a few students. But by now in this book we have crunched numbers using R through more complicated formulas. A square root or two is not going to scare us off.

From the box "Roots of a Quadratic Equation," the two roots of a quadratic equation, when they exist, are:

$$x_1 = \frac{-b - \sqrt{b^2 - 4ac}}{2a},$$

$$x_2 = \frac{-b + \sqrt{b^2 - 4ac}}{2a}.$$

The quantity under the square root symbol serves as a warning flag: if $b^2 - 4ac$ is a negative number, then no real number exists as its square root, and the roots x_1 and x_2 do not exist as real numbers. The situation occurs when the parabola fails to intersect the x-axis so that no real-valued roots to the equation $ax^2 + bx + c = 0$ exist (as in Figure 8.5b and d). If however $b^2 - 4ac$ is a positive number, then two real roots exist and are given by the above formulas. If $b^2 - 4ac$ equals zero, the values of x_1 and x_2 are identical, and the parabola just touches the x-axis at its high or low point.

Substituting $a = 1$, $b = -1$, and $c = -1$ in the quadratic root formulas gives us our formulas for the roots of the Golden Ratio quadratic equation:

$$x_1 = \frac{-(-1) - \sqrt{(-1)^2 - 4(1)(-1)}}{2(1)} = \frac{1 - \sqrt{5}}{2},$$

$$x_2 = \frac{-(-1) + \sqrt{(-1)^2 - 4(1)(-1)}}{2(1)} = \frac{1 + \sqrt{5}}{2}.$$

We can now finish the job of calculating the numerical value of the Golden Ratio. Go to the console and compute the two roots:

```
> x1=(1-sqrt(5))/2
> x2=(1+sqrt(5))/2
> x1
[1]-0.618034
> x2
[1] 1.618034
```

The second root is the Golden Ratio, to six decimal places. Mathematicians have proved that the square root of 5, and along with it the Golden Ratio, are irrational, and so any decimal representations are necessarily approximations. The root formulas themselves can be considered "exact" in the mathematical sense, but to design a building we would at some point need to perform calculations.

The value $x_2 - 1 \approx 0.618034$, the length of the short side of the smaller rectangle (Figure 8.2), is sometimes called the **Golden Section**. If you have a long thin stick and you want to break it into pieces that form a Golden Rectangle, first break it at about 61.8% of its length, then break each of the pieces in half.

Parabolas are symmetric, and the formulas for the roots identify the value of x where the high point or the low point occurs. The roots are in the form $-b/(2a)$ *plus something* and $-b/(2a)$ *minus something*. That point in the center (let us call it x^*), given by

$$x^* = -\frac{b}{2a},$$

is the value of x, where the parabola attains its highest or lowest point. The value of the quadratic function at the high or low point is found by evaluating the quadratic at x^*:

$$y^* = a\left(x^*\right)^2 + bx^* + c = a\left(\frac{-b}{2a}\right)^2 + b\left(\frac{-b}{2a}\right) + c = \frac{b^2}{4a} - \frac{b^2}{2a} + c = -\frac{b^2 - 4ac}{4a}.$$

The quantities x^* and y^* become important when a quadratic equation is used to model a relationship, which might have an optimum value. For instance, if y is the average yield of potato plants and x is the amount of nitrogen fertilizer applied, one might expect a hump-shaped dependence of y on x, as nitrogen fertilizer is beneficial to growth at low application levels but becomes toxic at high levels. Then, application of an amount x^* of fertilizer would produce the maximum growth. In such a use of a quadratic equation, the appropriate values of a, b, and c would be determined with data from agricultural experiments.

Real-World Example

Renewable resources are any economically valuable resources that grow and replace their losses. Various biological populations such as trees, fish, and shrimp possess biomass valuable to humans. The natural growth of the total biomass of those populations provides a harvestable resource that in principle can be replenished indefinitely. In the management of renewable resources, the concept of sustainable harvesting is central. When the biomass is harvested at a rate equal to the rate of biological growth, the harvest rate can be sustained indefinitely, provided conditions for growth do not change. The population will be in balance, neither increasing nor decreasing, at a level below that which it would attain without harvesting.

Managers of an ocean fishery estimate a quantity called fishing effort. Fishing effort can be thought of as the number of "boat days" of harvesting, that is, the number of standard fishing boats times the number of days of harvesting. Each boat in the fleet has different gear, and its actual effort contribution must be adjusted by complex calculations (the size of its nets, its average speed during harvesting, and so on).

Managers noticed decades ago that if the overall fishing effort is increased, the size of the sustainable harvest *per unit effort* tends to decrease. The size of the sustainable harvest is called the sustainable yield. When more boats go out to fish, the overall effort goes up, but the amount of fish harvested for each boat-day (in the long run) tends to decline. The situation occurs because the surplus biological production is divided up among more and more boat-days.

Suppose we denote the total fishing effort in standard boat-days by E. Also, we will denote the total sustainable yield of the harvest by Y. The yield-per-unit-effort becomes Y/E. The simplest model of how the yield-per-unit-effort decreases with effort is a line with a negative slope:

$$\frac{Y}{E} = h - gE.$$

Here the slope of the line is $-g$, with g positive. The sign of the slope is indicated explicitly to emphasize the decreasing nature of the relationship (a common practice in scientific studies). One, however, must be careful to keep track of the sign calculations. Multiplying each side by E gives the yield as a function of the total amount of harvesting effort:

$$Y = hE - gE^2.$$

As we can see, the function is a quadratic equation in which the coefficient (a) of the quadratic term is negative and the constant term (c) is zero. The function is hump-shaped, opening downward: yield rises initially with increasing effort, reaches a maximum, and then decreases to zero. The rightmost zero of the quadratic represents the point at which maximum biological growth of the resource is exceeded by the harvest rate represented by the total effort. Extinction of the resource occurs, and the long-run or sustainable yield is zero.

The Gulf of Mexico Fishery Management Council reported data from the Gulf of Mexico shrimp fishery for 1990 through 2005 (Ad Hoc Shrimp Effort Working Group 2006). A salient portion of the data is reproduced in Table 8.1.

TABLE 8.1

Variable "Yield" is the Total Landings of Shrimp (Millions of Pounds of Tails) for Each Year, and "Effort" is the Harvesting Effort (Thousands of Standard Boat-Days) from the Gulf of Mexico Shrimp Fishery for 1990 through 2005

year	yield	effort
1990	159.28	310.087
1991	144.81	301.492
1992	138.16	314.541
1993	128.43	288.132
1994	131.38	304.220
1995	145.62	254.281
1996	139.68	255.804
1997	131.26	291.778
1998	163.76	281.334
1999	150.87	270.000
2000	180.36	259.842
2001	159.97	277.777
2002	145.51	304.640
2003	159.87	254.598
2004	161.16	214.738
2005	134.30	150.019

Source: Ad Hoc Shrimp Effort Working Group, *Estimation of Effort, Maximum Sustainable Yield, and Maximum Economic Yield in the Shrimp Fishery of the Gulf of Mexico,* Report to the Gulf of Mexico Fishery Management Council, 2006.

Each row of data represents a year. The column labeled "yield" holds the total landings of shrimp (millions of pounds of tails) for each year, and the column labeled "effort" is the harvesting effort (thousands of standard boat-days) that year. A text file containing the data is posted at http://webpages.uidaho. edu/~brian/rsc/RStudentCompanionData.html for the typing averse.

Using a "curve-fitting" method, I calculated the values of *a* and *b* in the quadratic yield–effort relationship that provides the best-fitting parabola for the shrimp data. You will learn the curve-fitting method in Chapter 15 of this book. For now, all we need are the results:

$$g = 0.002866, \quad h = 1.342.$$

Let us plot the data in a scatterplot with an overlay of the fitted parabola. The task is well within our R skill set and will be an informative look at the management of an economically important renewable resource. Before proceeding further, you should try writing your own script to produce the graph. Then, compare it to the following script.

```
#==================================================================
# R script to draw a scatterplot of sustainable yield vs.
# harvesting effort for shrimp data from the Gulf of Mexico,
# with parabolic yield-effort curve superimposed.
#==================================================================

#------------------------------------------------------------------
# 1.  Read and plot the data.
#   (a data file named shrimp_yield_effort_data.txt is assumed
#    to be in the working directory of R)
#------------------------------------------------------------------
df = read.table("shrimp_yield_effort_data.txt",header=TRUE)
attach(df)
plot(effort,yield,type="p",xlab="harvest effort",
  ylab="sustainable yield")
detach(df)

#------------------------------------------------------------------
# 2.  Calculate parabolic yield-effort curve and overlay it on
#        scatterplot.
#------------------------------------------------------------------
elo=100              # Low value of effort for calculating
                     #    quadratic.
ehi=350              # High value of effort.
eff=elo+(0:100)*(ehi-elo)/100   # Range of effort values
g=0.002866
h=1.342
sy=-g*eff^2+h*eff    # Sustainable yield calculated for range of
                     #    effort values.
points(eff,sy,type="l")   # Add the quadratic model to the plot.
```

Before this script is run, the working directory of R should be changed to the folder on your computer where you have stored the text file with the shrimp data. Of course, instead of reading in the data from the text file, you could have alternatively just typed in the data into the vectors `yield` and `effort` using the `c()` statement. Go through the script and be sure you understand what every line does. The script is pretty tame; there are no R functions or R loops.

The resulting graph shows a simple quadratic yield–effort model but widely scattered data (Figure 8.6). The graph illustrates some of the difficulties of managing renewable resources. Biological systems are complex and driven by many known and unknown variables. The shrimp data conform to the model only in a vague way, but there is a hint in the data of the rising and falling shape of the quadratic. If that rising and falling pattern is accepted as real, then the graph suggests that the resource is somewhat overharvested. The quadratic model predicts that the management action of reducing the harvest would produce a rise in the average yield of the shrimp fishery. Go to the console and calculate the level of harvesting effort predicted to produce the maximum sustainable yield, along with the maximum sustainable

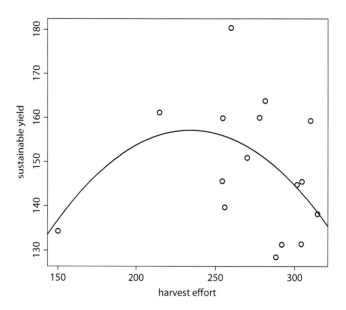

FIGURE 8.6

Relationship between harvest effort (thousands of boat-days) and sustainable yield (millions of pounds of tails landed) of shrimp in the Gulf of Mexico. Solid curve: fitted parabola of the form $Y = hE - gE^2$, with $h = 1.342$ and $g = .002866$, where Y is sustainable yield and h is the harvest rate. Circles: data from the Gulf of Mexico shrimp fishery for 1990 through 2005. (From Ad Hoc Shrimp Effort Working Group, *Estimation of Effort, Maximum Sustainable Yield, and Maximum Economic Yield in the Shrimp Fishery of the Gulf of Mexico*, Report to the Gulf of Mexico Fishery Management Council, 2006.)

yield itself. To use the formulas for x^* and y^* from above, perhaps it is clearest to define a, b, and c in terms of the quantities in the problem, then apply the formulas:

```
> a=-g
> b=h
> c=0
> eff.star=-b/(2*a)
> msy=-(b^2-4*a*c)/(4*a)
> eff.star
[1] 234.1242
> msy
[1] 157.0973
```

The maximum sustainable yield is predicted to occur around 234 thousand boat-days. However, the uncertainty arising from the poor fit of the model suggests that the prediction be treated with caution. More data are needed for the lower middle range of harvest efforts to obtain a clearer picture of the yield–effort relationship. Contemporary resource managers advocate the use of "adaptive management," which prescribes management policies in part as experiments to learn more about the system being managed. Of course, back in the year 2006, improved management toward the sustainable yield of shrimp in the Gulf of Mexico assumed that the resource would not be contaminated or destroyed by a catastrophic blowout of a Gulf oil well.

Final Remarks

It is difficult to overemphasize the importance of quadratic functions in applied mathematics and quantitative science. As one of the simplest functions with curvature, the quadratic function gets a lot of use as a mathematical model in all sorts of situations. In physics, we encounter a parabola as the trajectory of a projectile (say, a baseball) thrown into the air at an angle from the ground (Chapter 9). The shape of a common form of telescope mirror is a paraboloid (rotated parabola). The famous "bell-shaped curve" giving relative frequencies of heights, test scores, batting averages, and many other collections of quantities is a parabola when plotted on a logarithmic scale (Chapter 14). One of the early examples of mathematical "chaos" was demonstrated with a quadratic function (May 1974). A faulty economics model in the form of a parabola-like figure called the "Laffer curve" (originally sketched on a cocktail napkin by economist Arthur Laffer in 1974 in a bar near the White House, for the benefit of Dick Cheney, then deputy to White House Chief of Staff Don Rumsfeld) was responsible for the mistaken belief among "supply-side" conservatives in the United States during the years

1980–2010 that cutting taxes would lead to increased government revenues and thereby reduce deficits. The fault was not so much that the principle was wrong but rather that the absence of data points for the Laffer curve meant that policy makers did not actually know where the peak of the curve was. Quadratics are everywhere, for better or for worse.

ROOTS OF A QUADRATIC EQUATION

Formulas for the two points where a quadratic equation intersects the x-axis can be obtained algebraically. The method is called "completing the square." We start with the equation to be solved:

$$ax^2 + bx + c = 0.$$

By "solving" we mean finding any values of x, which when substituted into the above equation produces the identity $0 = 0$. The first step in solving is to subtract c from both sides:

$$ax^2 + bx = -c.$$

Divide both sides by a:

$$x^2 + \left(\frac{b}{a}\right)x = -\left(\frac{c}{a}\right).$$

The idea now is to add something to both sides that will make the left side a perfect square. That something is found by: (1) dividing the coefficient in the x term by 2 and (2) squaring the result:

$$x^2 + \left(\frac{b}{a}\right)x + \frac{b^2}{4a^2} = -\left(\frac{c}{a}\right) + \frac{b^2}{4a^2}.$$

We have completed the square: the left-hand side is the same as $\left(x + \frac{b}{2a}\right)^2$:

$$\left(x + \frac{b}{2a}\right)^2 = -\left(\frac{c}{a}\right) + \frac{b^2}{4a^2}.$$

We can get at x now by taking the square root of both sides. We can do this and obtain a real number as the result provided the right-hand side is positive. The right-hand side is positive when

$$-\left(\frac{c}{a}\right) + \frac{b^2}{4a^2} > 0$$

or (multiplying both sides by $4a^2$, which as a positive number does not alter the inequality)

$$b^2 - 4ac > 0.$$

When we take the square root, we must remember that two real numbers, one positive and one negative, can be squared to produce $\left(x + \dfrac{b}{2a}\right)^2$:

$$x + \frac{b}{2a} = -\sqrt{-\frac{c}{a} + \frac{b^2}{4a^2}} \quad \text{and} \quad x + \frac{b}{2a} = +\sqrt{-\frac{c}{a} + \frac{b^2}{4a^2}}.$$

Thus, we have found two different values of x that are solutions to the quadratic equation:

$$x_1 = -\frac{b}{2a} - \sqrt{-\frac{c}{a} + \frac{b^2}{4a^2}},$$

$$x_2 = -\frac{b}{2a} + \sqrt{-\frac{c}{a} + \frac{b^2}{4a^2}}.$$

The more familiar forms of the roots usually displayed in algebra books are algebraically equivalent and found by combining terms inside the square root radical using a common denominator:

$$x_1 = -\frac{b}{2a} - \sqrt{\frac{-4ac + b^2}{4a^2}} = \frac{-b - \sqrt{b^2 - 4ac}}{2a},$$

$$x_2 = -\frac{b}{2a} + \sqrt{\frac{-4ac + b^2}{4a^2}} = \frac{-b + \sqrt{b^2 - 4ac}}{2a}.$$

WHAT WE LEARNED

1. A quadratic function has the form $y = ax^2 + bx + c$, where a, b, and c are constants. The graph of a quadratic function is a curve called a parabola.

Example:

```
> a=-2
> b=4
> c=1
> x=-1+4*(0:100)/100
```

```
> y=a*x^2+b*x+c
> plot(x,y,type="l")
```

2. If $b^2 > 4ac$, the quadratic function given by $y = ax^2 + bx + c$ intersects the x-axis at two points called roots. The roots represent values of x for which $ax^2 + bx + c = 0$. The two roots are given by the quadratic root formulas.

Example:

```
> a=-2
> b=4
> c=1
> x1=(-b-sqrt(b^2-4*a*c))/(2*a)
> x2=(-b+sqrt(b^2-4*a*c))/(2*a)
> x1
[1] 2.224745
> x2
[1] -0.2247449
```

3. A special quadratic function results when $a = 1$, $b = -1$, and $c = -1$: $y = x^2 - x - 1$. One of the roots to this quadratic equation is the Golden Ratio, with value (to six decimal places) 1.618034. The Golden Ratio has been used in architecture as an aesthetically pleasing proportion for the lengths and heights of buildings.

Computational Challenges

8.1. Draw a graph of each of the four quadratic equations that appear in Figure 8.5:

   a. $y = x^2 - x - 1$
   b. $y = x^2 - x + 1$
   c. $y = -x^2 + x + 1$
   d. $y = -x^2 + x - 1$

8.2. Write a few R functions for calculating aspects of quadratic equations for your function collection:

   a. A simple R function to calculate the values of a quadratic equation for a vector of x values. The function arguments should be the three coefficients (a, b, c) as well as a vector containing the x values at which the quadratic is to be evaluated. The output shoud be a vector of y values.

b. An R function for calculating the roots of a quadratic equation, given any coefficients (*a, b, c*). The function arguments should be the coefficients (*a, b, c*). The function should check for whether the roots exist as real numbers. The function should have as output a vector containing the roots if they exist, and a warning text message if they do not exist.

c. The location of the maximum or minimum of a quadratic and its value. The function arguments should be the coefficients (*a, b, c*). The function output should be a vector containing the location of the maximum or minimum and the value of the quadratic at that maximum or minimum.

References

Ad Hoc Shrimp Effort Working Group. 2006. *Estimation of Effort, Maximum Sustainable Yield, and Maximum Economic Yield in the Shrimp Fishery of the Gulf of Mexico.* Report to the Gulf of Mexico Fishery Management Council.

May, R. M. 1974. Biological populations with nonoverlapping generations: Stable points, stable cycles, and chaos. *Science* 186:645–647.

9

Trigonometric Functions

Triangles are an abstract mathematical invention and have been used by human beings for many millennia (Figure 9.1). The mathematical properties of triangles have played a crucial role in human scientific and technological development, from the building of the pyramids to the modern understanding of the general theory of relativity and the universe.

Any three points not on a line determine a triangle, and triangles within their triangular essence have a large number of shapes. A general triangle has three sides and three interior angles. Geometry tells us that for every triangle the angular measures of the three angles always add up to 180°.

Right Triangles

One type of triangle has been singled out for its sheer usefulness. A triangle with one of its angles measuring 90° is a "right triangle." The angular measures of the other two angles of a right triangle must therefore add to 90°. The 90° angle of a right triangle is usually identified in pictures by a little box (Figure 9.2).

One of the immediately useful properties of a right triangle is given by the Pythagorean theorem. If the side opposite to the right angle, called the hypotenuse, has length r (such as in the large triangle in Figure 9.2) and the other two sides have lengths x and y, then for any right triangle the lengths are related as follows:

$$r^2 = x^2 + y^2.$$

The Pythagorean relationship can be used to calculate the length of the third side if lengths of the other two sides are known.

Another useful feature is that if two right triangles have an additional angle with the same measure, then they are "similar" triangles: Their side lengths have the same proportions. For example, for the two triangles shown in Figure 9.2 all the corresponding ratios of side lengths are equal: $y/x = w/v$, $x/r = v/q$, and so on.

FIGURE 9.1
An Egyptian papyrus, dated 1650 BCE, photographed in the British Museum, London, United Kingdom, by the author. Could it be some ancient student's lost trigonometry homework? All that math looks like hieroglyphics to me.

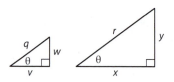

FIGURE 9.2
Two right triangles with additional angles of identical measure θ. The triangles are therefore similar and have sides with lengths in the same proportions, for instance, $w/v = y/x$.

Trigonometric Functions

The usefulness of the similarity property of right triangles is that once one side length and one additional angle of a right triangle are measured, the lengths of the other sides can be calculated. This idea is the basis for "triangulating" distances from a baseline or line of known length. The idea has proved so useful through the centuries, in surveying land, designing buildings, calculating ranges for artillery, and estimating distances to stars, that the ratios of side lengths of right triangles were calculated and cataloged in large tables. These ratios of side lengths of right triangles comprise the "trigonometric functions." They are functions of the angular measure θ of an additional angle of the right triangle. Once θ is known, the ratios are fixed in value from the similarity property. For an angle with angular measure θ in a right triangle, there are six possible ratios for side lengths. These six ratios define the six basic trigonometric functions: (1) sine, (2) cosine, (3) tangent, (4) cotangent, (5) secant, and (6) cosecant (usually abbreviated sin, cos, tan, cot, sec, and csc, respectively). The functions defined for the larger triangle shown in Figure 9.2 are as follows:

$$\sin \theta = \frac{y}{r} \quad \csc \theta = \frac{r}{y},$$

$$\cos \theta = \frac{x}{r} \quad \sec \theta = \frac{r}{x},$$

$$\tan \theta = \frac{y}{x} \quad \cot \theta = \frac{x}{y}.$$

The trigonometric functions appearing on the same line in the above expressions are reciprocals of each other. Because of these reciprocal relationships, you can do most trigonometric calculations using just sin, cos, and tan. The popular (and weirdly suggestive) mnemonic for remembering the big three trigonometric functions is SOH CAH TOA (i.e., sine: opposite over hypotenuse; cosine: adjacent over hypotenuse; tangent: opposite over adjacent).

R has the built-in functions `sin()`, `cos()`, and `tan()`. However, to use them, you must learn to think in radians rather than degrees.

Right Triangles, Circles, and Radians

Right triangles and circles are intimately bound. Take a Cartesian coordinate system and draw a circle centered at the origin with radius r. Pick any point on the circle in the positive quadrant; suppose it has coordinates (x, y) as shown in Figure 9.3. Draw a line segment from the point to the origin; this segment is the hypotenuse of a right triangle with length $r = \sqrt{x^2 + y^2}$. The side with length y is formed by a line segment drawn in a vertical direction to the horizontal axis, with the right angle being formed by the junction of the vertical line segment and the horizontal axis. Let θ be the measure of the angle formed by the hypotenuse segment and the horizontal axis (measured from the positive horizontal axis to the hypotenuse segment in a counterclockwise direction). Now, θ can have any value from near 0° to near 90°, and all the values are possible for a right triangle. All the points (x, y) in the positive quadrant that are at a fixed distance r from the origin form a quarter arc of the circle. Every possible shape of a right triangle is represented by that quarter arc.

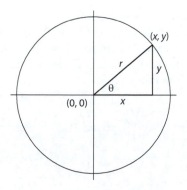

FIGURE 9.3

A right triangle with a hypotenuse of length r inscribed in a circle of radius r.

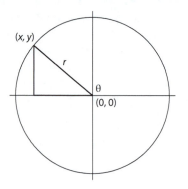

FIGURE 9.4

A right triangle with a hypotenuse of length r inscribed in a circle of radius r. For an obtuse angle θ, use coordinates x and y (possibly negative numbers) in the trigonometric function definitions, not just lengths. Use r as a length.

Next, pick a point on the circle in some other quadrant (Figure 9.4). Again draw a line segment from the point to the origin. The angle between the positive horizontal axis and the line segment measured in the counterclockwise direction is "obtuse," that is, θ is greater than 90° (angles less than 90° are said to be acute). The right triangle in the positive quadrant is gone. However, another right triangle is formed in the new quadrant, using a vertical line segment drawn from the point on the circle to the horizontal axis. The definitions of trigonometric functions are extended to angles greater than 90° by the following conventions: All the points on the circle define all values for θ between 0° and 360° by measuring the angle from the positive horizontal axis to the line segment in a counterclockwise direction. The trigonometric functions for the obtuse values of θ retain their definitions as above, except that x and y are explicitly coordinates (not just lengths) with possibly negative values. The value r is always a hypotenuse length and is always positive. For instance, if θ is 135° (putting the point on the circle in the north-west quadrant) the value of x is negative and the cosine, tangent, secant, and cotangent of θ are all negative.

The circumference of a circle is $2\pi r$. If you walk counterclockwise around a circle of radius $r = 1$, the distance you have walked on completing the whole circle is 2π. The connection between circles and right triangles leads to the natural notion of measuring angles in terms of distance. In this sense, a degree is not an angle measurement explicitly connected to distances on a plane. Applied mathematics has almost universally adopted a more conve-nient angle measurement based on the arc lengths of a circle.

Imagine that you start at the positive horizontal axis at the point (1, 0) and walk counterclockwise all the way around a circle of radius $r = 1$ while hold-ing a string of length 1 joined to the origin. The angle formed by the string and the positive horizontal axis is said to have traversed 2π **radians** (instead of 360°). The angle measure in radians at any point along the journey is the distance you have walked on the circle. When you finish one-quarter of the way around you have traveled a distance of $\pi/2$ and, so, your string-angle measures $\pi/2$ radians (instead of 90°). A distance of π corresponds to 180°:

Degrees	Radians
0	0
45	$\pi/4$
90	$\pi/2$
135	$(3\pi)/4$
180	π
225	$(5\pi)/4$
270	$(3\pi)/2$
315	$(7\pi)/4$
360	2π

In order to extend the idea, think of going around the circle more than once to just add more distance to your walk. The angle traversed by your string on completion of two times around the circle, for instance, will be 4π radians. Thus, the angle measurement of radians is extended in the positive direction to the entire positive real line, when the angle is formed by going around more than once. Of course, the string will be in the same position if, for instance, you traverse $\pi/4$ radians (45°) or $(9/4)\pi$ radians (360° + 45° = 405°). You can envision that the basic properties of the triangle formed by you, your string, and the horizontal axis, such as ratios of side lengths, are just going to repeat themselves as you go round and round the circle.

Finally, to complete the idea, think of starting at (1, 0) and walking around the circle in the clockwise direction. Mathematics considers this to be accumulating angular measure in a negative direction. Thus, if you travel over a clockwise distance $\pi/4$, the angular measurement of your string-angle is said to be $-\pi/4$. The angular measurement is in this way extended to the negative portion of the real line; angles can be of positive or negative radians based on the direction in which they are measured and on how many times around the circle they are measured.

An angle measurement is thereby a *directed* measurement expressed in units of radians and can take any real-number value. The concept of measuring an angle in this generalized way is incredibly useful and has paved the way for huge advances in scientific understanding. The reason is that nature is full of things that repeat themselves, such as orbits, waves, pendulums, and bouncing springs. The key idea is to map the repeating phenomena of nature to the repeating properties of right triangles while time goes round and round a circle. The idea was a milestone breakthrough in technological progress.

The point is, science thinks in radians, and now you should too. R, in particular, thinks only in radians.

The main trigonometric functions in R are sin(), cos(), and tan(). The arguments are assumed to be radians and can be vectors. A trigonometric function in R with a vector argument transforms every element according to the function and returns the results in a vector. Let us try some examples. Remember that pi in R is a reserved word that returns the value of π to many decimal places. Go to the console (at last!) and type the following R commands:

```
> theta=c(0,(1/4)*pi,(2/4)*pi,(3/4)*pi,pi,(5/4)*pi,(6/4)*pi,
   (7/4)*pi,2*pi)
> sin(theta)
[1]   0.000000e+00   7.071068e-01   1.000000e+00   7.071068e-01
[5]   1.224647e-16  -7.071068e-01  -1.000000e+00  -7.071068e-01
[9]  -2.449294e-16
> cos(theta)
[1]   1.000000e+00   7.071068e-01   6.123234e-17  -7.071068e-01
[5]  -1.000000e+00  -7.071068e-01  -1.836970e-16   7.071068e-01
[9]   1.000000e+00
> tan(theta)
[1]   0.000000e+00   1.000000e+00   1.633124e+16  -1.000000e+00
[5]  -1.224647e-16   1.000000e+00   5.443746e+15  -1.000000e+00
[9]  -2.449294e-16
```

The values in the vector theta go around the circle in amounts of $\pi/4$ radians (or 45°). We can see in the results printed on the console that trigonometric functions have some recurring values and negatively reflected values. We can also see that the values are not quite exact; for instance, we know from the definition of the sine function that $\sin(2\pi) = 0$ but R tells us that it is $-2.449294 \times 10^{-16}$. The value given by R is indeed very, very small; it is close enough to zero so as not to matter in *most* calculations. Such round-off errors are almost always present in computer calculations.

Let us try a graph and get a better picture of the functions in our minds. Open the R editor and enter the following script:

```
th.lo=-4*pi
th.hi=4*pi
theta=th.lo+(th.hi-th.lo)*(0:1000)/1000    # Values of theta
                                           # ranging
                                           # from -4*pi to
                                           # +4*pi.
y1=sin(theta)
y2=cos(theta)
plot(theta,y1,type="l",lty=1,ylim=c(-2,2),xlab="theta",
  ylab="sine and cosine")
points(theta,y2,type="l",lty=2)
```

The script produces plots of the sine and cosine functions, as shown in Figure 9.5.

We see that sin θ and cos θ appear as periodic waves. Now try a script for the tangent function:

```
th.lo=-4*pi
th.hi=4*pi
theta=th.lo+(th.hi-th.lo)*(0:1000)/1000    # Values of theta
                                           # ranging
                                           # from -4*pi to
                                           # +4*pi.
y=tan(theta)
plot(theta,y,type="p",ylim=c(-2,2),xlab="theta",
  ylab="tangent")
```

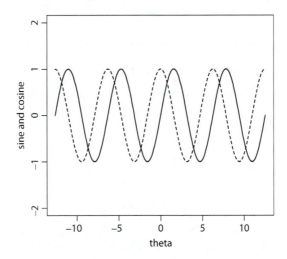

FIGURE 9.5

Plots of sin θ (solid curve) and cos θ (dashed curve) functions, with values of θ in radians.

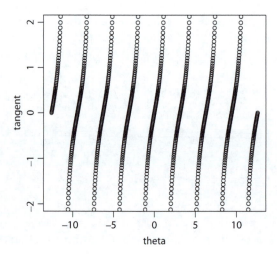

FIGURE 9.6
Values of tan θ, using values of θ in radians.

This little script produces Figure 9.6. In the figure, the tangent function was drawn with points instead of a line so that the ends of the separate curves would not be connected with each other. From the definition of tangent, that is, tan θ = y/x, we see that the function becomes infinite or negatively infinite as the value of x approaches 0. The value of x is 0 when θ is a positive or negative odd-integer multiple of $\pi/2$ (90°): ..., $-5\pi/2, -3\pi/2, -\pi/2, \pi/2, 3\pi/2, 5\pi/2, ...$. Whether the function goes to a negative or positive infinite value depends on whether x is approaching 0 from the positive side or the negative side.

Properties of Trigonometric Functions

From the definitions of trigonometric functions as ratios, we can see various simple relationships among the functions, for example,

$$\sin \theta = \frac{1}{\csc \theta},$$

$$\tan \theta = \frac{\sin \theta}{\cos \theta},$$

and so on. Such simple relationships among trigonometric functions lead to a bewildering variety of algebraic identities. If (when) you take a trigonometry class, you will experience the whole parade of identities and formulas in their vast glory. We will mention just a couple of formulas here.

First, look at Figure 9.3. From the figure and the Pythagorean theorem, we can see that the equation for a circle is

$$x^2 + y^2 = r^2,$$

that is, the set of all points (x, y) that satisfy the equation constitute a circle of radius r centered at the origin. Divide both sides by r^2 to get

$$\left(\frac{x}{r}\right)^2 + \left(\frac{y}{r}\right)^2 = 1.$$

By the definition of sine and cosine, the above equation is

$$(\sin \theta)^2 + (\cos \theta)^2 = 1.$$

This famous trigonometric identity is just the Pythagorean theorem expressed in terms of trigonometric functions.

A second famous trigonometric formula is the general triangle formula. Draw a triangle like the one given in Figure 9.3, but alter the right angle to any angle permissible in a triangle (i.e., it must be strictly less than π or 180°), so that the triangle is no longer a right triangle. Suppose the measure of this interior angle is ϕ (Figure 9.7). Divide the triangle into two right triangles, each with height h (Figure 9.7). (If either angle on the horizontal axis is obtuse, i.e., greater than $\pi/2$, then draw a right triangle outside the triangle in question using a vertical line from the high vertex to the horizontal axis; the proof of the general triangle formula follows steps similar to the ones mentioned here.) The two right triangles each have their own Pythagorean relationships:

$$r^2 = h^2 + (x-a)^2,$$

$$y^2 = h^2 + a^2.$$

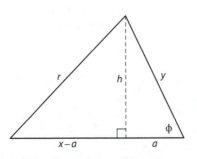

FIGURE 9.7
Combining Pythagorean relationships for two right triangles. Combining the relationships produces the general result $x^2 + y^2 - 2xy \cos \phi = r^2$, which is valid for all angles of all triangles.

The algebraic steps are as follows: Solve both equations for h^2, equate the two expressions, substitute $x^2 - 2xa + a^2$ for $(x - a)^2$, and substitute $y \cos \phi$ for a (from the definition $\cos \phi = a/y$) to get the following general triangle formula:

$$x^2 + y^2 - 2xy \cos \phi = r^2.$$

Try the algebraic steps; they are not hard to understand. The formula states that in a triangle, the squares of two adjacent sides of an angle, minus a correction that depends on how much the angle departs from a right angle, add up to the square of the side opposite to the angle. It is a generalization of the Pythagorean theorem that applies to all angles of all triangles. Taking $\phi = \pi/2$ produces the ordinary Pythagorean result.

Polar Coordinates

In Cartesian coordinates, as shown in Figure 9.3, each point on a plane is represented by an ordered pair of real numbers (x, y). Figure 9.3 illustrates that any such point can also be represented by a different ordered pair of real numbers comprising the distance r from the origin and the measure θ of the angle between the positive horizontal axis and the line segment between the origin and (x, y). Here, the distance r must be nonnegative and the angle θ is between 0 and 2π, inclusive.

The ordered numbers (r, θ) are called "polar coordinates" of the point (x, y). Polar coordinates help simplify many mathematical derivations, such as, for example, planetary orbits in physics. Given the polar coordinates (r, θ) of a point, one obtains the corresponding Cartesian coordinates by applying simple trigonometric calculations:

$$x = r \cos \theta,$$

$$y = r \sin \theta.$$

Going the other way, from Cartesian coordinates to polar coordinates, is easy for getting the distance r due to the following Pythagorean result:

$$r = \sqrt{x^2 + y^2}.$$

Getting θ from x and y using a formula is a bit more difficult, and such a formula is not presented here. From the definition $\tan \theta = y/x$, we see that the tangent function has to be "undone" somehow to get θ. The undoing of trigonometric functions involves the use of "inverse trigonometric functions," a topic that is a little too involved (not difficult, only somewhat long) to develop in this book. The inverse trigonometric functions are often denoted as arcsin, arccos, and so on. The main complication is that for the inverse tangent function one needs a different formula for θ for each quadrant.

For us, in this chapter, a good use of polar coordinates is drawing circles and other forms that loop around some sort of a center. For instance, to draw a circle of radius r centered at the origin, take a range of values of θ from 0 to 2π and calculate vectors of x and y values for plotting. An R script for the purpose might look like this:

```
theta=2*pi*(0:100)/100
r=1
x=r*cos(theta)
y=r*sin(theta)
par(pin=c(4,4))
plot(x,y,type="l",lty=1)
```

The `par()` statement is an R function for global plotting options, and the `pin=c(4,4)` arguments set the plotting region to be 4×4 in. The reason for using this option is that the default actual computer screen distances on the x- and y- axes are not equal. A circle plotted without using this option will look somewhat elongated on the computer screen. The script produces Figure 9.8.

You can make some fun shapes by having r change with θ. For instance, you can let r increase as a linear function of θ, which gives a spiral (Figure 9.9); this was first invented by the ancient Greek scientist Archimedes:

```
theta=10*pi*(0:1000)/1000   # Go 5 times around & use many
                            # points.
r=.1+theta/(2*pi)   # Radius increases 1 unit each time around.
x=r*cos(theta)
y=r*sin(theta)
par(pin=c(4,4))
plot(x,y,type="l",lty=1)
```

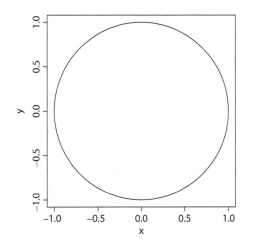

FIGURE 9.8
Circle graphed with the method of polar coordinates.

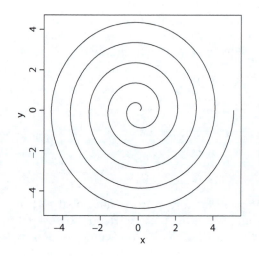

FIGURE 9.9
Archimedes spiral.

There are many variations of polar curves. Look on the Internet and you can discover the "Polar Rose," "Lemniscate of Bernoulli," "Limaçon of Pascal," and others. I call the following one "Archimedes with an iPod":

```
theta=10*pi*(0:1000)/1000
r=.1+theta/(2*pi)+.5*sin(10*theta)
x=r*cos(theta)
y=r*sin(theta)
par(pin=c(4,4))
plot(x,y,type="l",lty=1)
```

Triangulation of Distances

From geometry, two angles and a side determine a triangle (the AAS property). In Figure 9.3, we know that the angle measure of one of the angles is $\pi/2$ (or 90°). If the angle measure θ is known and one of the distances is known, then the lengths of the other two sides of the triangle can be calculated. For instance, if y is the unknown height of a tree and the distance x to the base of the tree and the angle measure θ sighted to the top of the tree are determined, then the height of the tree is given by

$$y = x \tan \theta,$$

from the definition of tangent.

Real-World Examples

Distances to Stars Near the Solar System

Distances to stars near our solar system can be accurately measured by tri-angulation using what astronomers call **parallax**. Astronomers measure the angle to a star when Earth is on one side of its orbit and then measure the angle again 6 months later when Earth is on the opposite side. If we denote the amount by which the angle has changed by 2θ (half the amount of change θ is called the parallax), we can construct a right triangle as shown in Figure 9.10.

Using 1 AU as the (average) distance from the Earth to the sun, the equation for the distance from the sun to the star is distance = $1/\tan\theta$. The angle represented by θ in Figure 9.10 for the star Proxima Centauri, which is the closest star to the sun, is 3.7276×10^{-6} radians. The distance in light years (LYs) can be obtained from the fact that 1 LY is about 63279 AU (light travels from the sun to the Earth in a little over 8 minutes). A small script serves to find the distance to the nearest star:

```
theta=3.7276e-06        # Parallax.
dist.au=1/tan(theta)    # Distance in AU.
dist.au                 #   --
[1] 268269.1            #   --
dist.ly=dist.au/63270   # 63270 LY per AU.
dist.ly                 # Distance in LY.
[1]  4.240068           #   --
```

Light takes over 4 years to travel to us from Proxima Centauri. Likewise, any intelligent life near the star who is monitoring our television signals is finding out only now (in 2013) who won *American Idol* in 2009.

Parallax angles for stars farther away than about 100 LYs are too small to be measured accurately from the Earth's surface. Satellites launched from Earth are extending parallax-determined distances to beyond 1000 LYs.

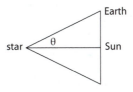

FIGURE 9.10
Shifts in the position of a nearby star by an angular amount of 2θ as seen from the Earth in observations spaced 6 months apart. The angular measure θ is the parallax. The distance from the sun to the star is $1/\tan\theta$ AU.

Projectile Motion

In Chapter 1, Computational Challenge 1.4, some equations were given for the motion of a baseball thrown at an angle of 45° at an initial velocity of 75 mph. The equations for the motion of a baseball thrown at any (forward) angle and any (low) initial velocity contain some trigonometric functions. Let x be the horizontal distance traveled and let y be the vertical distance traveled, and both are measured in meters (to frame the problem in universal scientific units). If you throw a baseball at an angle of θ radians at an initial velocity of v_0 meters per second, the ball's initial velocity in the x direction is $v_0 \cos \theta$ and the initial velocity in the y direction is $v_0 \sin \theta$ (picture these using a rectangle having a diagonal length v_0 and the quantities $v_0 \cos \theta$ and $v_0 \sin \theta$ as the lengths of the sides).

If the ball is thrown while standing on a level field, the horizontal distance x traveled by the ball after t seconds is described (neglecting air resistance) by the following equation from Newtonian physics:

$$x = (v_0 \cos \theta)t.$$

Furthermore, the height of the ball above the ground after t seconds, assuming it was initially released at a height of y_0 meters, is described by

$$y = y_0 + (v_0 \sin \theta)t - \frac{g}{2}t^2,$$

where g is the gravity acceleration constant ($g \approx 9.81$ m/s²). We recognize the equation for x to be a linear function of t and the equation for y to be a quadratic function of t.

The ball hits the ground when $y = 0$. The time t_{max} at which this happens is thus the larger root of a quadratic equation:

$$t_{max} = \frac{-b - \sqrt{b^2 - 4ac}}{2a},$$

where $a = -g/2$, $b = v_0 \sin \theta$, and $c = y_0$. The farthest distance traveled before hitting the ground is then $x_{max} = (v_0 \cos \theta)t_{max}$ from the equation for x.

We can gather all these equations into a handy script for calculating and plotting the motion of projectiles in general:

```
#================================================================
# R script to calculate and graph projectile motion (such as
# throwing a baseball).
#================================================================
#----------------------------------------------------------------
# Input initial velocity, angle, and height, in USA common units.
#----------------------------------------------------------------
```

```
mph=75              # Initial velocity, miles per hour.
angle=45            # Initial angle, degrees.
height=5            # Initial height, feet.


#-----------------------------------------------------------------
# Convert units to meters and seconds.
#-----------------------------------------------------------------
v0=mph*1609.344/(60*60)  # Convert velocity to meters per second.
theta=2*pi*angle/360     # Convert angle to radians.
y0=height/3.2808399      # Convert height to meters.
g=9.80665                # Gravitational acceleration constant,
                         #   meters per second per second.


#-----------------------------------------------------------------
# Calculate maximum time of flight using quadratic root formula.
#-----------------------------------------------------------------
a=-g/2
b=v0*sin(theta)
c=y0
t.max=(-b-sqrt(b^2-4*a*c))/(2*a)   # Max. time of flight.
x.max=v0*cos(theta)*t.max          # Max. distance


#-----------------------------------------------------------------
# Plot height at time t vs distance at time t.
#-----------------------------------------------------------------
t=t.max*(0:50)/50   # Range of t values between 0 and t.max.
x=v0*cos(theta)*t
y=y0+v0*sin(theta)*t-g*t^2/2
plot(x,y,xlab="distance in meters",ylab="height in meters")
                                               # Plot.
t.max                                          # Print t.max.
```

The first part of the script obtains data such as the initial values of velocity, angle, and height in the measurement units of miles per hour, degrees, and feet, respectively, which are units commonly used to discuss baseball in the United States. You are welcome to change the script if you can think in the scientifically preferred units of meters per second, radians, and meters. If you are a U.S. baseball fan, you are welcome to reconvert the distance units to feet before plotting. The script outputs the plot in meters and produces Figure 9.11. The ball thrown at 45° at 75 mph from an elevation of 5 ft on a level field will fly for about 4.9 seconds and travel about 116 meters (381 ft). The circles in Figure 9.11 are separated by time intervals of 4.9/50 = 0.098 seconds. Such a throw from home plate can clear the home run wall in many baseball stadiums. A more realistic analysis would take into account air resistance and the peculiar drag patterns that spinning baseballs generate.

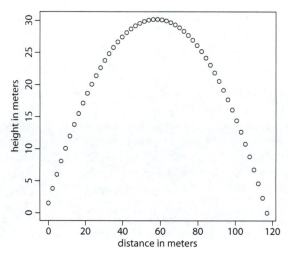

FIGURE 9.11
Circles showing positions of a projectile at equal time intervals after launching.

Planetary Orbits

Newton's law of gravity can be used to derive a polar curve equation for the orbit of a planet around the sun or the orbit of a satellite around a planet:

$$r = r_0 \frac{1 + \varepsilon}{1 + \varepsilon \cos \theta}.$$

The sun or body being orbited is assumed to be the origin. Here, r_0 is the distance of the point of closest approach (the **periapsis**) of the two bodies, and ε is the **eccentricity** (departure from circularity) of the orbit. The periapsis and eccentricity both depend on the initial velocity and direction of the movement of the orbiting body in complicated ways. For Earth's orbit around the sun, r_0 is about 0.98329 AU and ε is about 0.016711. With an R script, let us take a look and gaze down on the sun–Earth system from far away:

```
r0=0.98329
eps=.016711
theta=10*pi*(0:1000)/1000
r=r0*(1+eps)/(1+eps*cos(theta))
x=r*cos(theta)
y=r*sin(theta)
par(pin=c(4,4))
plot(x,y,type="l",lty=1,xlim=c(-1.1,1.1),ylim=c(-1.1,1.1))
```

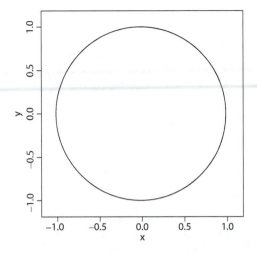

FIGURE 9.12
Earth's orbit around the sun plotted in a plane.

The xlim= and ylim= options set both the horizontal axis and the vertical axis limits identically at –1.1 and +1.1 so that the shape of the orbit is not distorted by different scales. The script produces Figure 9.12.

As can be seen in Figure 9.12, the Earth's orbit is nearly circular. Other orbits in the solar system, such as those of Pluto and the comets, are substantially noncircular. In the orbit equation, a value $\varepsilon = 0$ for the eccentricity corresponds to a perfect circle, whereas $0 < \varepsilon < 1$ produces an ellipse. If $\varepsilon = 1$, the trajectory is not a closed loop but a parabola. The body is traveling so fast that it is not captured by the sun; instead it executes a parabolic flyby. When $\varepsilon > 1$, as in cases of extreme velocity, the flyby is a hyperbola.

In one of the amusingly stoic moments in Jules Verne's *From the Earth to the Moon* (1865), the adventurers, after taking note that their craft (shot out of a big cannon) is going to miss the moon and fly off into outer space forever, debate whether their trajectory past the moon will be parabolic or hyperbolic.

Final Remarks

The derivation of the orbit equation from Newton's law of gravity is a substantial calculus problem and is beyond our reach in this book. However, we have at our disposal now some powerful computational means. In Chapter 16, we start with Newton's laws and produce Earth's orbit using sheer numerical brute force.

WHAT WE LEARNED

1. The trigonometric functions are ratios of the sides of a right triangle. If the measure of one of the acute angles in the triangle is θ, then

$$\sin \theta = \frac{y}{r} \quad \csc \theta = \frac{r}{y},$$

$$\cos \theta = \frac{x}{r} \quad \sec \theta = \frac{r}{x},$$

$$\tan \theta = \frac{y}{x} \quad \cot \theta = \frac{x}{y}.$$

Here, r is the length of the hypotenuse, x is the length of the side adjacent to the angle θ, y is the length of the side opposite to the angle θ, and the functions are (when reading down the columns) sine, cosine, tangent, cosecant, secant, and cotangent.

2. An alternative measure for angles is radians. If the vertex of an angle is at the center of a circle of radius 1 unit, the angle measured in radians is the distance traversed counterclockwise along the arc of the circle inside the angle. Going around 360° is the same as going around 2π radians (the circumference of the circle). An angle of 90° measures $\pi/2$ radians.

3. Generalized radians can be greater than 2π (going around the circle more than once) or negative (going around clockwise). The values of trigonometric functions defined as various ratios of distance r and coordinates x, y repeat themselves each time around the circle.

4. The trigonometric functions in R are sin(), cos(), tan(), csc(), sec(), and cot(). All the functions take angle measurements in radians. The functions also take vector arguments and calculate the values of the function for each element.

5. The point (x, y) in a Cartesian coordinate system has an alternative representation in polar coordinates as the distance r of the point from the origin and the angle θ formed by the segment connecting the point to the origin and the positive horizontal axis. Given polar coordinates (r, θ), one can calculate the Cartesian coordinates as follows:

$$x = r \cos \theta,$$

$$y = r \sin \theta.$$

6. When the angle measure θ of one of the acute angles in a right triangle and the length of one of the sides are known, the other side lengths can be calculated using trigonometric relationships (triangulation), by using the definitions of trigonometric functions. For instance, if θ and x are known, then $y = x \tan \theta$.

7. Using the distance from the Earth to the sun as a known distance and the angular change of positions of a star in the sky in observations spaced 6 months apart (Earth at opposite sides of the sun), the distance to the star can be calculated directly using triangulation. The angles are large enough to measure only nearby stars.

8. The orbit of a planet around the sun plotted in a plane is a simple polar curve. The curve can be derived from Newton's gravitational laws.

Computational Challenges

9.1. Parallax angles (θ in Figure 9.10) for some nearby stars are given here in arcseconds. It is noted that 1 arcsecond is 4.848137×10^{-6} radians. Calculate their distances from the sun in AU. Convert these distances to LYs (1 LY ≈ 63,270 AU). There are bonus points for vectorizing the whole set of calculations.

star	parallax (arcseconds)
Alpha Centauri	.747
Sirius	.379
61 Cygni	.286
Procyon	.286
Vega	.130
Arcturus	.089
Capella	.077
Aldebaran	.050
Betelgeuse	.0051
Antares	.0059
Polaris	.0075
Rigel	.0042

9.2. From the following table of eccentricities and periapses, draw a plot of the orbit of each planet:

planet	ε	r_0
Mercury	.2056	.3075
Venus	.006773	.7184
Earth	.01671	.9833

Mars	.09341	1.381
Jupiter	.04839	4.951
Saturn	.05415	9.203
Uranus	.04717	18.28
Neptune	.008586	29.81
(Pluto)	.2488	29.66

Afternotes

With the convention of defining the circle constant π as the ratio of the circle's circumference to its diameter, mathematical formulas became filled with the term 2π. An interesting movement is afoot to change the circle circumference measure to "tau" defined by $\tau = 2\pi = 6.283 \ldots$. As the ratio of a circle's circumference to its radius (instead of diameter), τ is sort of connected in a more fundamental way to polar coordinates, trigonometry, and other mathematical concepts, and with its use a lot of formulas are simplified. As is common these days, such crusades have websites and online videos. Search on the Internet for "pi is wrong," "tau," and so on to get a feel for the arguments.

10

Exponential and Logarithmic Functions

Achieving Real Power

You have likely seen the square root function $\sqrt{x}$ often represented as $x^{1/2}$ (raising x to the 1/2 power), the cube root function $\sqrt[3]{x}$ as $x^{1/3}$ (raising x to the 1/3 power), and so on. The notation highlights the idea that the algebraic properties of exponents extend to roots. Similar to $x^1 x^1 = x^{1+1} = x^2$, we can write $x^{1/2} x^{1/2} = x^{(1/2)+(1/2)} = x$. Further, $x^{2/3}$ can be interpreted as either taking the cube root of x and squaring the result or squaring x and taking the cube root of the result: $x^{2/3} = (x^{1/3})^2 = (x^2)^{1/3}$.

In general, mathematics defines "raising x to a rational power m/n" or $x^{m/n}$, where m and n are positive integers and x is a positive real number, as the nth root of x^m or, equivalently, $x^{1/n}$ raised to the mth power. Further, an exponent with a negative sign denotes reciprocal, that is, $x^{-m/m} = 1/x^{m/n}$.

The right-hand side of the "raising to a power" operator "^" in R works with any rational number or vectors of rational numbers as the argument. All floating point numbers stored in R are terminating decimal numbers, so, strictly speaking, the numerical operations in R are performed only with rational numbers. Often the result is an approximation due to round-off error. Play with the power operator a bit at the R console and get familiar with the operator and the properties of exponents:

```
> x=10
> sqrt(x)
[1] 3.162278
> x^(1/2)
[1] 3.162278
> x^2*x^2
[1] 10000
> x^(3/2)
[1] 31.62278
> (x^3)^(1/2)
[1] 31.62278
> (x^(0:6))^(1/2)
[1]     1.000000      3.162278     10.000000     31.622777
[5]   100.000000    316.227766   1000.000000
```

If a is a rational number, we can draw a graph of x^a for a range of values of a, fixing the value of x and thinking of x^a as a function of a. The following is a little script to this end using the fixed value $x = 2$:

```
a=4*(0:100)/100-2   # Range of values of a between -2 and +2.
x=2                 # Value of x fixed at 2.
y=x^a
plot(a,y,type="l")
```

The resulting graph is shown in Figure 10.1. It is a rising curve that has height between 0 and 1 when $a < 0$ and greater than 1 when $a > 0$. However, the function x^a drawn in this way by connecting the dots with a line gives the impression that the function is defined on all real values of a. How would we interpret x^a when a is an irrational number? When a is not a ratio m/n of integers, there is no simple sequence of operations such as "raise x to the mth power and then take the nth root."

But mathematicians know the real-number line to be "dense" with rational numbers, that is, any irrational number can be approximated to any arbitrary precision by a rational number. One merely uses as many decimal places of the irrational number as are needed for the task at hand, leaving off the remainder of the infinitely continuing decimal digits. We know the value of π to millions of decimal places, although we rarely need more than a handful. So we can adopt a working definition of x^a, where a is any real number, as simply $x^{m/n}$, where m and n are integers chosen so that m/n is as close to a in value as our machines allow. We then consider the curve in Figure 10.1

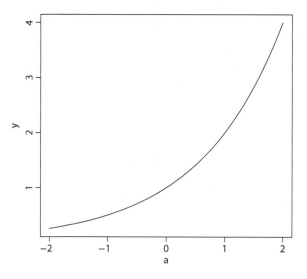

FIGURE 10.1
Graph of the function $y = x^a$ for varying values of a, with the value of x fixed at 2.

for all practical computing purposes as "smooth," with no gaps for irrational values of the exponent. We thereby achieve real power.

Raising x to a real power obeys all the algebraic exponent laws you learned for integer powers:

$$x^0 = 1,$$

$$x^1 = x,$$

$$x^{-u} = 1/x^u,$$

$$x^u x^v = x^{u+v},$$

$$(x^u)^v = x^{uv}.$$

In the above expressions, u and v are real numbers and x is any positive real number. Further, in mathematics 0^0 is usually defined to be 1, because the definition completes certain formulas in a sensible fashion.

Try out your newly found powers at the R console. The square root of 2 and π can serve as examples of numbers known to be irrational. Remember that in R "pi" is a reserved name that returns the value of π to double precision:

```
> 0^0
[1] 1
> pi
[1] 3.141593
> x=2
> x^pi
[1] 8.824978
> (x^pi) * (x^pi)
[1] 77.88023
> x^(pi+pi)
[1] 77.88023
> x^sqrt(2)
[1] 2.665144
```

The Special Number *e*

Take a look at the following power function:

$$y = \left(1 + \frac{1}{x}\right)^x.$$

When the value of x is large, the term $1/x$ inside the parentheses is small and the whole quantity in parentheses is close to the value 1. However, a large value of *x* also means that the exponent is large. The quantity in parentheses gets smaller as *x* gets bigger, but at the same time the quantity gets raised to a higher and higher power. As *x* becomes bigger, the function is pulled in two different ways. Will the function increase or decrease? The battle is on; who will be the winner?

Curious? One thing about R is that its ease of use invites numerical experiments. It is a cinch to calculate values of the function for a range of increasing values of *x*. The number of commands needed just to calculate the function is small, so we could easily do it at the console. However, let us do this calculation as a script and use a few extra commands to display the results nicely in a table. Let us throw in a graph of the function as well:

```
x=1:50              # Range of x values from 1 to 50.
y=(1+1/x)^x         # Function values.
function.table=data.frame(x,y) # Put x and y in two columns.
function.table                 # Print to the console.
plot(x,y,type="l")   # Plot the function.
```

A long table of numbers printed at the console and the graph in Figure 10.2 are the results of running the above script. The function looks like it is leveling off, somewhere around the value 2.7.

But is the function actually leveling off? Close the graph, and try changing the first statement in the script to read as follows:

```
x=1:1000
```

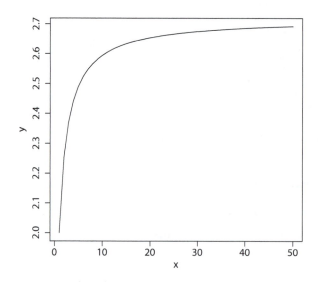

FIGURE 10.2
A graph of the function $y = \left(1+\dfrac{1}{x}\right)^x$. The graph becomes more and more level as *x* becomes large.

Rerun the script and view the resulting plot. Although the figure is not mathematical proof, the figure does leave a strong impression that the function levels off somewhere slightly above 2.7.

"Levels off" is an imprecise way of describing the pattern. Actually, the function is always increasing (this can be proved using a little calculus) as x becomes larger. However, there is an upper level that the function will approach closer and closer in value but will never cross. The function approaches what is termed in mathematics a "limit." The limit of this function is a famous real number called e.

The number e is known to be irrational. Its value to a few decimal places is

$$e = 2.71828\ldots.$$

Interestingly, the value e emerges from the expression $\left(1 + \dfrac{1}{x}\right)^x$ when x is negative and gets smaller (goes toward minus infinity). Try it:

```
x=-(1:1000)      # Range of x values from -1 to -1000.
y=(1+1/x)^x      # Function values.
function.table=data.frame(x,y) # Put x and y in two columns.
function.table                 # Print to the console.
plot(x,y,type="l")   # Plot the function.
```

This time, the function decreases as x becomes more and more negative (going in the left direction), approaching closer and closer to e (Figure 10.3).

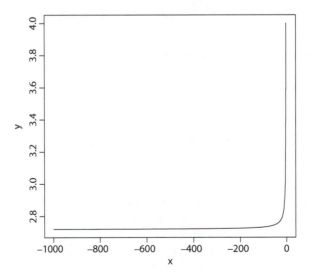

FIGURE 10.3
A graph of the function $y = \left(1 + \dfrac{1}{x}\right)^x$. The graph becomes more and more level as x becomes more and more negative.

The Number *e* in Applications

The number *e* crops up over and over in all kinds of applications in applied mathematics. One reason is that the power function $\left(1+\dfrac{1}{x}\right)^{x}$ from which *e* springs as a limit occurs frequently in mathematical modeling as a crucial step when going from discrete time ($t = 0, 1, 2, 3, ...$) to continuous time ($0 < t$, where *t* is real valued). For instance, think of investing \$1 at a 100% interest rate (we can dream, can't we?) for 1 year. Remember the investment function, that is, $y = m(1+r)^{t}$, where *y* is the amount after *t* years, *r* is the interest rate, and *m* is the initial amount invested. After 1 year, the amount of money we would have is \$2:

$$y = (1)(1+1)^{1} = 2.$$

Now, suppose the interest is paid after each quarter year. The quarterly interest rate is 1/4, and there are 4 time units in the projection:

$$y = (1)\left(1+\frac{1}{4}\right)^{4} \cong 2.441406.$$

The numerical evaluation on the right-hand side is done with a quick referral to the R console. Now, suppose the interest is paid monthly. The monthly interest rate will be 1/12, with 12 time units until a year is reached:

$$y = (1)\left(1+\frac{1}{12}\right)^{12} \cong 2.613035.$$

If interest is paid daily,

$$y = (1)\left(1+\frac{1}{365}\right)^{365} \cong 2.714567.$$

You can see where this is going, can't you? In general, as the year gets chopped up into finer and finer pieces, with time becoming more continuous and less discrete, \$1 at 100% annual interest compounded continuously will yield close to *e* dollars rather than \$2 at the end of a year.

A slight adjustment gives us continuous compounding for any annual interest rate. Suppose the annual interest rate is represented by *r* and there are *w* interest payments in 1 year. Each interest payment would occur at the rate of *r/w*, and the total amount of money after 1 year, for each dollar invested, would be

$$y = (1)\left(1 + \frac{r}{w}\right)^{w}.$$

Rewrite r/w as $1/(w/r)$ and rewrite the exponent w as $(w/r)r$:

$$y = (1)\left(1 + \frac{1}{(w/r)}\right)^{(w/r)r}.$$

Write $u = w/r$, and you can see that as w gets bigger (finer and finer time divisions), u also gets bigger. We have

$$y = (1)\left(1 + \frac{1}{(w/r)}\right)^{(w/r)r} = (1)\left(1 + \frac{1}{u}\right)^{ur} = (1)\left[\left(1 + \frac{1}{u}\right)^{u}\right]^{r} \to e^{r}.$$

The arrow expresses the concept that the quantity on the left gets closer and closer to the quantity on the right. The number e raised to the power r gives the number of dollars resulting when \$1 is invested at an annual interest rate of r with the interest compounded continuously. The quantity on the right approximates the quantity on the left when the number of interest payment periods during 1 year is very large. The quantity e^{r} results in the derivation as well when r is negative, for example, when there is a penalty interest continuously subtracted from the account. In such a case, the quantity $u = w/r$ is negative in the above approximation formulas and goes toward minus infinity instead of infinity (as in Figure 10.3).

Raising e to some power is called the "exponential function," and it is a hugely useful calculation in many aspects of quantitative sciences. We will now learn to take advantage of the exponential function in R.

The Exponential Function

The exponential function is defined by

$$y = e^{x},$$

where x is any real number. The resulting value of y is a positive real number between 0 and 1 when x is negative and greater than 1 when x is positive. In mathematical formulas, when x in turn is represented as a complicated calculation or function involving other quantities so that the tiny exponent notation might be difficult to read because of additional subscripts or superscripts, the exponential function is often denoted as

$$y = \exp(x).$$

For instance, $\exp\left(-\dfrac{z^2}{2}\right)$ can be used instead of $e^{-\frac{z^2}{2}}$.

Similar to trigonometric functions, the exponential function is a special preprogrammed function in R. The syntax is exp(x), where x is a quantity or a vector of quantities. If x is a vector, the exponential function in R acts on each element of x and returns a corresponding vector of values. Go to the console and try it out:

```
> x=1
> exp(x)
[1] 2.718282
```

Raising e to the power 1 produces the (computer-approximate) value of e. Try more examples:

```
> x=0:10
> exp(x)
 [1]    1.000000    2.718282    7.389056   20.085537
 [5]   54.598150  148.413159  403.428793 1096.633158
 [9] 2980.957987 8103.083928 22026.465795
> x=-(0:10)
> exp(x)
 [1] 1.000000e+00 3.678794e-01 1.353353e-01 4.978707e-02
 [5] 1.831564e-02 6.737947e-03 2.478752e-03 9.118820e-04
 [9] 3.354626e-04 1.234098e-04 4.539993e-05
```

Remember that 3.678794e-01 in computer scientific notation means 3.678794×10^{-1}, which is the same as 0.3678794. The "e" in computer scientific notation stands for exponent, and not the number e.

Exponential Growth

We saw in a previous section (The Number e in Applications) that \$1, invested at an annual interest rate r compounded continuously (e.g., daily, to an approximation) yields e^r dollars after 1 year. After 2 years, there would be $e^r e^r = e^{2r}$ dollars, after 3 years there would be $e^r e^r e^r = e^{3r}$ dollars, and so on. Evidently after t years the initial \$1 would become $(e^r)^t = e^{rt}$ dollars. Here, t can be a real number too so that one can calculate the dollars after, say, 3.56 years.

If instead of just one dollar there are *m* dollars initially and if we let *n* denote the number of dollars after time *t*, we arrive at the equation of exponential growth in continuous time:

$$n = me^{rt}.$$

In the exponential growth equation, me^r is the number of dollars in the investment after 1 year. For positive values of *r*, e^r is greater than 1 and, so, me^r is greater than *m*. The money increases. However, some investments lose money. Money-losing investments can often be modeled with the exponential growth equation in which the value of *r* is negative. When *r* is less than 0, e^r is less than 1 and me^r is less than *m*. The money decreases.

The exponential growth equation is fundamentally important due to its many uses in science and business. In biology, it is the model of a population's growth after *m* individuals (or units of biomass) colonize a new resource. The model might be expected to be a good description of the population only for the initial years, though, before resources for growth (i.e., for producing more individuals or biomass) run out. In physics, the exponential growth equation with a negative value of *r* is used to describe the "exponential decay" of an initial amount *m* of a radioactive substance. Here, *n* is the amount of the radioactive substance left after time *t*. We look further into these concepts in the section "Real-World Examples."

Let us draw some pictures of exponential growth. Consider three curves on one figure, with *r* values of 0.04, 0, and −0.04 and an initial amount *m* of 5. A small script does the trick:

```
# R script to plot exponential growth for positive, zero, and
# negative growth rates.
m=5         # Initial size of exponentially changing quantity.
t=0:20      # values of time from 0 to 20.
r=.04
n=m*exp(r*t)
plot(t,n,type="l",ylim=c(0,12)) # Vertical axis limits from 0
                                #  to 12.
r=-.04
n=m*exp(r*t)
points(t,n,type="l")
r=0
n=m*exp(r*t)
points(t,n,type="l",lty="dashed") # Zero growth rate plotted with
                                  #   dashed line.
```

The script when run produces Figure 10.4. The vertical axis limits were picked by trial and error after a few runs to get a nice-looking graph. The upper curve illustrates continuous exponential increase, whereas the lower

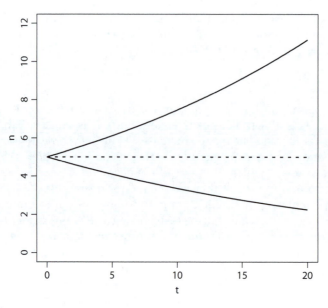

FIGURE 10.4
Plots of the exponential growth function given by $n = me^{rt}$ for $r = 0.04$ (increasing curve), $r = 0$ (level line), and $r = -0.04$ (decreasing curve), using initial condition of $m = 5$.

curve depicts exponential decay. The dashed horizontal line is the result when $r = 0$.

Logarithmic Functions

The exponential function $y = e^x$ is a function of x that is always increasing as x increases (Figure 10.5a). Imagine the function is drawn on a transparency and that you pick up the transparency and turn it over, rotating it so that x is depicted as a function of y (Figure 10.5b). The function we are looking at now is the **logarithmic function**. Whatever power to which you raise e to get y, that power is the "natural logarithm" of y and it is written as follows:

$$x = \log(y).$$

One way of writing the logarithm definition is as follows:

$$y = e^{\log(y)}.$$

Logarithms were invented more than 400 years ago in part as a way of reducing multiplication and division (complex numerical tasks) to addition (a relatively simple numerical task) by taking advantage of the "adding exponents" rule. If $y = e^u$ and $z = e^v$, then $yz = e^{u+v}$. Evidently $u + v = \log(yz)$; but

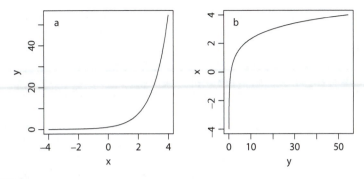

FIGURE 10.5
Figure showing two graphs: (a) graph of the exponential function given by $y = e^x$ and (b) graph of the (base e) logarithmic function given by $x = \log(y)$.

because $u = \log(y)$ and $v = \log(z)$, we see that the logarithm of a product is the sum of the logarithms of the numbers multiplied:

$$\log(yz) = \log(y) + \log(z).$$

Also, $y/z = e^u/e^v = e^{u-v}$, so the logarithm of a quotient is the difference between the logarithms of the numbers undergoing division:

$$\log(y/z) = \log(y) - \log(z).$$

The multiplication/division procedure would involve looking up $\log(y)$ and $\log(z)$ in a big table of numbers and their logarithms (just a detailed table of the logarithmic function shown in Figure 10.5b), adding/subtracting the logarithms, and then doing a reverse lookup in the table to find the number whose logarithm corresponds to the sum/difference.

My engineer grandfather's slide rule is on my desktop monitor. A slide rule is a simple pair of rulers with etched logarithmic (base 10 instead of base e) scales. A skilled slide rule user (are there any left?) can quickly get most multiplications or divisions to three significant digits.

Any positive number besides e can serve as the base for a system of logarithms. Logarithms with a base of 10 are taught in many precalculus mathematics courses and are encountered in some science and engineering applications (in particular, in the definition of pH in chemistry and the Richter earthquake magnitude scale in geology, as described in the Logarithmic Scales section). We denote "log to the base 10" of a positive real number y as $\log_{10}(y)$, with the following definition:

$$y = 10^{\log_{10}(y)}.$$

In theoretical computer science and information theory, one encounters base 2 logarithms, which are denoted as $\log_2(y)$, for analyzing binary (0, 1) data representations.

Some precalculus mathematics books just use the notation "log( )" to denote base 10 logarithms, whereas some calculus books use the notation "ln( )" to denote base e logarithms (natural logarithms) to avoid confusion. The number e pops up in so many calculus formulas for scientific applications that anyone working in the sciences will sooner or later start using natural logarithms most of the time.

The point of this long discussion is that in R the natural logarithm function is log(). If you have an urgent need for a base 10 logarithm, you can find it in R as log10(). Also, base 2 logarithms are precoded in R as log2().

Too much talk and not enough R. Let us try some logarithms:

```
> w=0
> log(w)
[1] -Inf
> w=1:10
> log(w)
 [1] 0.0000000 0.6931472 1.0986123 1.3862944 1.6094379
 [6] 1.7917595 1.9459101 2.0794415 2.1972246 2.3025851
> w=10000
> log10(w)
[1] 4
> w=16
> log2(w)
[1] 4
```

The logarithmic function is handy in algebra. The function is used to "get at" quantities that are stuck in exponents. Because $y = e^x$ means $x = \log(y)$, we can think of taking logarithms of both sides of $y = e^x$ in order to bring down x:

$$y = e^x,$$

$$\log(y) = \log(e^x) = x.$$

The exponential and logarithmic functions are inverse functions of each other. Inverse here does not mean one is the reciprocal of the other. Rather, it means one erases the other's doings:

$$\log(e^x) = x,$$

$$e^{\log(y)} = y,$$

where x is a real number and y is a positive real number. Raising e to a power x is sometimes called "taking the antilogarithm of x" and is accessed on some scientific calculators as "inv-log" or "inv-ln."

Also, if a is a positive real number, then $a = e^{\log(a)}$ and, so, taking the logarithm of a^x brings down x multiplied by $\log(a)$:

$$a^x = e^{[\log(a)]x},$$

$$\log(a^x) = [\log(a)]x.$$

The above formula is the key to going back and forth between logarithms in base e and other bases. If

$$y = a^x = a^{\log_a(y)} = e^{[\log(a)]x},$$

then

$$\log(y) = [\log(a)][\log_a(y)],$$

and so the log to any base a expressed in terms of base e logarithms is

$$\log_a(y) = [\log(y)]/[\log(a)].$$

In R, a function for $\log_a(y)$ is log(y,a).

Logarithmic Scales

In science, some phenomena are measured for convenience on a logarithmic scale. A logarithmic scale might be used for a quantity that has an enormous range of values or that varies multiplicatively.

Richter Scale

A well-known example of a logarithmic scale is the Richter scale for measuring the magnitude of earthquakes. The word magnitude gives a clue that the scale is logarithmic. Richter magnitude is defined as the base 10 logarithm of the amplitude of the quake waves recorded by a seismograph (amplitude is the distance of departures of the seismograph needle from its central reference point). Each whole number increase in magnitude represents a quake with waves measuring 10 times greater. A magnitude 6 quake has waves that measure 10 times greater than those of a magnitude 5 quake. The offshore earthquake that sent a devastating tsunami to the northeast shore of Japan in March 2011 had a Richter magnitude of 9. The largest quake ever recorded is the enormous 1960 quake in Chile of magnitude 9.5, tsunamis from which killed people in places as far away as Japan. The quake that destroyed San Francisco, California, in 1906 is estimated to have been a magnitude 8.0 earthquake.

The pH Scale

Another logarithmic scale is pH, which is used in chemistry. The pH of an aqueous (water-based) solution is an inverse measure of the strength of its acidity: As pH decreases, the acidity of the solution increases. An aqueous solution is water with one or more dissolved chemicals. Some chemicals alter the degree to which hydrogen ions (protons) separate from water molecules. The hydrogen ions are highly reactive and want to rip apart and combine with just about any other molecules they encounter; hence, strong acids are able to burn or dissolve many substances with which they come in contact. In a student's first pass through chemistry, pH is usually defined as minus the base 10 logarithm of the concentration (moles per liter) of hydrogen ions in a solution, or $-\log_{10}([H^+])$, where $[H^+]$ denotes the concentration (moles/liter) of hydrogen ions. The letters pH stand for "powers of hydrogen." In later chemistry courses, for exacting calculations one must use a more complex definition of pH that takes into account the fact that different substances in a solution can measurably change the activity of hydrogen ions in the solution.

Pure water has a pH of 7, representing a tiny amount of water molecules (each with two hydrogen atoms and one oxygen) that naturally dissociates into a hydrogen ion and an oxygen–hydrogen ion (hydroxyl ion; denoted by OH^-). Solutions with pH less than 7 are called acids. Some substances dissolved in water accept and bind with hydrogen ions, leading to pH values greater than 7. These solutions are called "bases." Because hydrogen ions are "soaked up," bases have an excess of hydroxyl ions, which are also highly reactive. Strong bases can dissolve things similar to strong acids. A few substances and their approximate pH values (given in parentheses) are battery acid (0), sulfuric acid (1), lemon juice (2), soda (3), tomato juice (4), black coffee (5), milk (6), sea water (8), baking soda (9), milk of magnesia (10), ammonia solution (11), soapy water (12), oven cleaner (13), and liquid drain cleaner (14). There can be substances with negative pH values, as presumably was the case for the circulatory fluid of the monster in the sci-fi film *Alien*.

Star Magnitude

The ancient Greeks classified visible stars into six categories depending on their brightness in the night sky. This categorization of stars according to their apparent brightness as seen from the Earth persists in modern-day astronomy. First magnitude stars are the brightest ones, many with familiar names: Sirius, Alpha Centauri, Betelgeuse, Rigel, Vega, and so on. The North

Star, Polaris, is a run-of-the-mill magnitude 2 star, as are six of the main seven stars in the Big Dipper. On a superbly clear dark night, away from city lights, good eyes can see magnitude 6 stars (the night sky under such conditions is a tremendous sight to behold and is increasingly elusive due to the spread of light pollution). Millions of stars of magnitude 7 and dimmer can be viewed through binoculars and telescopes.

In the 1800s, when apparent brightness could be measured with some precision, astronomers devised a logarithmic brightness scale that retained the spirit of the ancient six categories. It was noted that an average magnitude 1 star was about 100 times brighter than a magnitude 6 star. What kind of logarithmic base did this system have? Suppose we take a magnitude 5 star to have x times the brightness of a magnitude 6 star, a magnitude 4 star to have x times the brightness of a magnitude 5 star, and so on. Then, if b is the brightness of a magnitude 6 star, $100b$ is the brightness of a magnitude 1 star; we see that

$$100b = xxxxxb = x^5b.$$

After canceling all the bs, we get

$$100 = x^5.$$

Take the fifth root of both sides of the equation (raise both sides to the 1/5 power) to get

$$x = 100^{1/5} = \sqrt[5]{100}.$$

R gets us the value of the base of the fifth root of 100, which is the base of the logarithmic scale of apparent star brightness:

```
> 100^(1/5)
[1] 2.511886
```

So, each whole number decrease in star magnitude represents around a 2.5-fold increase in apparent star brightness. Some reference sky objects and their magnitudes are Venus (maximum brightness –4.89), Sirius (–1.47), Alpha Centauri (–0.27), Vega (0.03), Antares (1.03), and Polaris (1.97), and the bowl stars of the Little Dipper counting clockwise from the brightest have magnitudes close to 2, 3, 5, and 4. The number $100^{1/5}$ is called "Pogson's ratio" after the astronomer Pogson who was principally responsible for its invention and adoption as the base of a scale for measuring star magnitude.

Brightness is called **luminosity** by astronomers. If L is the luminosity of a star as seen from the Earth and M is its magnitude, then the two quantities are related as

$$L = 100^{-M/5}.$$

The minus sign denotes luminosity decreases as magnitude increases. Solving for M (try this) produces magnitude as a function of luminosity:

$$M = -5\frac{\log(L)}{\log(100)}.$$

Real-World Examples

We will derive, analyze, and graph three mathematical models of real-world phenomena for which an understanding of exponential and logarithmic functions is greatly helpful.

Radioactive Decay

Suppose we single out an atom that is potentially subject to atomic decay. For instance, an atom of carbon-14 has six protons and eight neutrons in its nucleus. A neutron is a particle comprising a proton and an electron bound together by nuclear forces. At any time, one of the neutrons can spontaneously "pop" and emit its electron, which flies away at a tremendous speed. The carbon-14 atom is thus transformed into a nitrogen-14 atom with seven protons and seven neutrons. The carbon-14 atom is said to have "decayed" into a nitrogen-14 atom. For intricate reasons, as understood by atomic physics today, some combinations of protons and neutrons are more stable than others. Nitrogen-14 and carbon-12 (six protons and six neutrons), for instance, are highly stable.

Suppose we take an amount of time t and divide it into many tiny intervals, each having length h. The number of such intervals would be t/h. The laws of quantum physics indicate the chance that a carbon-14 atom decays is the same in every little time interval. If h is quite small, we can represent this chance as λh, where λ is a positive constant that has a characteristic value for carbon-14 but different values for different types of unstable atoms. A rapidly decaying atom has a high value of λ, and a more stable atom has a low value of λ. The chance that the atom does not decay during the tiny interval is $1 - \lambda h$. Because there are t/h such intervals, the chance that the atom does not decay in time t is $1 - \lambda h$ raised to the power t/h:

Probability that the atom does not decay in time t
$$= (1 - \lambda h)^{t/h} = \left[1 + \frac{1}{-1/(\lambda h)}\right]^{-\lambda t/(-\lambda h)} = \left[\left(1 + \frac{1}{u}\right)^{u}\right]^{-\lambda t} \rightarrow e^{-\lambda t}.$$

Here, $u = -1/(\lambda h)$, which goes toward minus infinity as h becomes small. Thus, an exponential function given by $e^{-\lambda t}$ represents the probability that a given atom will not decay within time t.

Suppose there are m atoms, a number likely to be quite enormous even for a speck of matter, initially. Then, $e^{-\lambda t}$ represents to a close approximation the fraction of atoms that have not decayed by time t. The fact is embodied in a result from probability called the "law of large numbers," which is discussed in Chapter 14. Flip a fair coin a billion times; the fraction of times that heads occurs will be very, very close to 1/2.

Denote by n the number of atoms left undecayed at time t. The above argument leads us to the following mathematical model for n:

$$n = me^{-\lambda t}.$$

This model for **exponential decay** gives the number of atoms remaining "unpopped" as an exponentially decreasing function of time.

One concept that is frequently referred to in connection with radioactive decay is the **half-life** of the radioactive substance. The half-life is the amount of time taken for half of the atoms to decay (with half of the atoms remaining undecayed). For a fixed value of λ, we can use a little bit of exponential function algebra to solve for the amount of time needed for only half of the atoms to remain undecayed. In the exponential decay model, set $n = m/2$ (half of the initial number of atoms) and then solve for the value of t that makes it so:

$$\frac{m}{2} = me^{-\lambda t}.$$

Divide both sides by m:

$$\frac{1}{2} = e^{-\lambda t}.$$

Time t is stuck in the exponent. How to get at it? Take logarithms of both sides of the equation:

$$\log\left(\frac{1}{2}\right) = \log\left(e^{-\lambda t}\right) = -\lambda t.$$

Divide both sides by $-\lambda$ and the resulting value of t is the half-life of the substance. We will denote half-life by $t_{1/2}$:

$$t_{1/2} = \frac{\log\left(\frac{1}{2}\right)}{-\lambda} = \frac{\log(1) - \log(2)}{-\lambda} = \frac{\log(2)}{\lambda}.$$

The resulting formulas

$$t_{1/2} = \frac{\log(2)}{\lambda},$$

$$\lambda = \frac{\log(2)}{t_{1/2}},$$

allow one to calculate the half-life $t_{1/2}$ from the decay rate λ, and vice versa. Carbon-14, for instance, has a half-life of around 5730 years. Its decay rate is easily calculated in R:

```
> t.half=5730
> lambda=log(2)/t.half
> lambda
[1]  0.0001209681
```

Carbon-14 and other unstable radionuclides are used for dating objects from the near and distant past. In the case of carbon-14, the dating technique is based on the fact that carbon-14 is formed at a roughly constant rate in the upper atmosphere by cosmic rays hitting nitrogen atoms. The resulting balance between formation and decay of carbon-14 causes the ratio of unstable carbon-14 atoms to stable carbon-12 and stable carbon-13 atoms in the atmosphere to stay near an equilibrium level of around 1 part per trillion (600 billion carbon-14 atoms per mole of carbon). Plants take in carbon dioxide from the atmosphere and fix it in their tissues by photosynthesis, a process that is (essentially) indiscriminate among carbon-12, -13, or -14 atoms. The proportion of carbon-14 atoms in the carbon atoms fixed by a plant into, say, cellulose is close to the atmospheric proportion. However, carbon-14 is not replenished in the cellulose once the cellulose is made, and the number of carbon-14 atoms starts to decrease according to the exponential decay equation. That is, a radioactive clock starts ticking.

Suppose we have an ancient piece of wood from an archaeological site. Modern laboratories can measure the number of carbon-14 atoms in a small sample of the wood using sensitive decay particle counters. Suppose a sample of the wood contains only 223 billion carbon-14 atoms per mole of carbon. We use m in the equation as the initial number of carbon-14 atoms in a mole of the sample we would expect to find when the wood is freshly produced, and n is the number of carbon-14 atoms per mole of the sample as revealed by the laboratory particle counters. In the equation, we then know the values of m, n, and λ. All we need to do is algebraically solve the equation for t and then substitute the numerical values of m, n, and λ for our particular case; thus we will get the approximate number of years for which the wood has existed after its first growth. Try some exponential function algebra. Divide both sides of the exponential decay equation by m, take logarithms of both sides, and then rearrange the terms to get the following equation:

$$t = \frac{\log(n/m)}{-\lambda}.$$

Now you can use the R console to crunch the numbers:

```
> n=223
> m=600
> t.half=5730
> lambda=log(2)/t.half
> t=log(n/m)/(-lambda)
> t
[1] 8181.975
```

We estimate that the wood was first grown around 8182 years ago. The simple, uncalibrated carbon-14 calculation illustrated here is not bad, but it is now known to be inaccurate by a few hundred years. The carbon-14 dating technique has been refined considerably since its invention around 1950. Many small adjustments are made now to calibrate for small fluctuations in atmospheric carbon-14 proportion (due mostly to variations in sunspot activity) and the small but measurable differences in the rates with which plants fix carbon-14, carbon-13, and carbon-12. The decay rate of carbon-14 is high enough that too little carbon-14 remains for useful dating of a carbon-containing object beyond about 60,000 years. Other radionuclides with much longer half-lives are used for dating older items. The accuracy of dating with carbon-14 and other radionuclides has been independently verified with tree ring series, lake bed sediment layers, growth rings in corals, and other methods. Macdougall (2008) has written an excellent introduction to the history and uses of radiometric dating in the sciences.

The R statements above for calculating the decay time *t* repeated the calculation for λ and used assignment statements for defining *n* and *m*, instead of just throwing the numbers themselves into the equation for *t*. This highlights the idea that the calculation might more usefully be written in general, for any half-life of any radionuclide and any data. You can anticipate a computational challenge at the end of the chapter of transforming the console statements above into a useful R script or function.

Limit to Population Growth

The equation of exponential population growth reflects the fact that organisms increase in number multiplicatively. Just like each dollar in a bank account adds a few pennies to the account each year, each individual in a population *on average* adds some small amount of new individuals to the population each year. Of course, individuals in a population likely vary dramatically in their reproductive output, that is, a few females may give birth, whereas males, many females, and juveniles may contribute to no actual additions to the population in a given year, or even represent negative contributions in the event of deaths. The sense in which the word average is used

here is that the net births and deaths *per individual* acts for the population like
the interest rate in a bank account.

An important property of exponential growth is revealed by taking logarithms of both sides of the exponential growth equation:

$$n = me^{rt},$$

$$\log(n) = \log(me^{rt}) = \log(m) + \log(e^{rt}),$$

or

$$\log(n) = \log(m) + rt.$$

The right-hand side of the result is seen to be a linear function of time t,
with the vertical axis intercept being $\log(m)$ (logarithm of the initial population size) and the slope being r. If r is positive, the net births in the population are greater than the net deaths and the logarithm of population size
increases linearly through time. If r is negative, the logarithm of population
size decreases linearly through time. Population ecologists frequently plot
the logarithm of a population's size through time, when linear trends in the
plot reveal episodes of exponential increase or decrease.

However, populations cannot continue to grow indefinitely. As populations become more crowded, resources run out, disease spreads more easily,
and competition for survival takes energy from reproduction. Populations
grow by transforming some limiting resource into biomass. Antelopes transform grass, wolves transform moose, finches transform seeds, and so on. In
some cases, such as trees and mussels, the limiting resource can be space,
which individuals must have in order to feed themselves and reproduce.

Population ecologists often model the growth of populations in the presence of a limiting resource in the following way. Suppose we denote k as
the maximum population size that an environment can sustain indefinitely.
As before, n is the population size at time t and m is the initial population size (at $t = 0$). The quantity $k - n$ is the potential additional population
size that can be added; it is a surrogate for the amount of unused resource
that remains in the environment. The ratio $(k - n)/n$ of unused resource to
used resource might be expected to decrease through time. Its initial value
is $(k - m)/m$. For reasons related to the multiplicative nature of population
growth, the revised population model formulates logarithm of the ratio
of unused resource to used resource, given by $\log[(k - n)/n]$, as a linearly
decreasing function of time with a starting point of $\log[(k - m)/m]$:

$$\log\left[\frac{k-n}{n}\right] = \log\left[\frac{k-m}{m}\right] - bt.$$

In this equation, b is a positive number and, so, $-b$ represents a negative slope
for the linear function of time. This revised model is universally called the

"logistic model" of population growth, and it was invented by the French mathematician Verhulst (1838).

Our goal is to draw a graph of population size over time, according to the logistic model. The equation will be handier if we solve it for n as a function of t so that we can graph population size over time. The solving process will give our new found algebraic powers a workout.

In order to get at n, which is inside a logarithm function, we need to take antilogarithms of both sides of the equation:

$$\exp\left[\log\left(\frac{k-n}{n}\right)\right] = \exp\left[\log\left(\frac{k-m}{m}\right) - bt\right] = \exp\left[\log\left(\frac{k-m}{m}\right)\right]\exp[-bt].$$

On the right, we have invoked the adding-exponents property. Everywhere $\exp\left[\log(\ )\right]$ occurs, the two functions wipe themselves out, leaving

$$\frac{k-n}{n} = \frac{k-m}{m}\, e^{-bt}.$$

On the left, $(k-n)/n$ is the same as $(k/n) - 1$. Add 1 to both sides of the equation, take reciprocals of both sides, and multiply both sides by k to get the sought-after expression for n as a function of t:

$$n = \frac{k}{1 + \left(\dfrac{k-m}{m}\right)e^{-bt}}.$$

We are now in a position to do some graphing in R. Let us fix values of k, b, and m and then calculate n for a range of values of t. Remember that k is the upper maximum value of population size (if the initial population size m is below k) and b determines how fast the population size approaches k. We can add a dashed line to the graph to indicate where k is in relation to the population size. The following is the required script:

```
#===============================================================
# R script to plot the logistic model of population growth.
#===============================================================
k=100       # Population levels off at k.
b=.5        # Higher value of b will result in quicker approach
            # to k.
m=5         # Initial population size is m.
t=(0:100)*20/100   # Range of values of time t from 0 to 20.
n=k/(1+((k-m)/m)*exp(-b*t))   # Logistic population growth
                              # function.
plot(t,n,type="l",xlab="time",ylab="population size") # Plot
                                                      # n vs t.
k.level=numeric(length(t))+k     # Vector with elements all
                                 # equal to k.
points(t,k.level,type="l",lty=2) # Add dashed line at the
                                 # level of k.
```

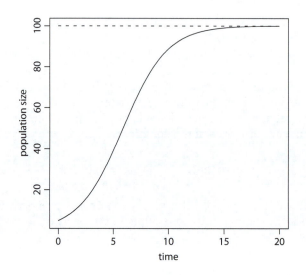

FIGURE 10.6
Figure showing two curves: solid curve shows graph of the logistic function as a model of population growth and the dashed line shows maximum population size according to the model.

The script produces Figure 10.6. The logistic function predicts that the population size will follow an S-shaped curve that approaches closer and closer to the value of k as time t becomes large. Notice that the population size appears to increase at nearly an exponential rate at first and then the rate of growth during each time unit slows down. The point where the curve stops taking a left turn and starts taking a right turn is called an "inflection point," and such curves are often said to have a "sigmoid" shape.

Peak Oil

In the logistic population model, a biological population replaces some underlying resource, nutrient, or substrate with more biological population. In the process, the underlying resource becomes depleted. In this sense, the logistic model could be used in many other contexts as a general model of something replacing something else. Some examples are biological invasions, such as cheat grass replacing native grass in western North American rangelands; epidemics, with infected individuals replacing susceptible individuals; commerce, such as Walmart replacing Kmart; innovation, such as cell phones replacing landline phones and digital versatile disc (DVD) replacing the home video system (HVS); human performance, with Fosbury high jump technique replacing frontal technique; and social ideas, with acceptance of interracial marriage replacing rejection of interracial marriage. The logistic model basically summarizes many processes in which one quantity loses market share to another.

Following such reasoning, the logistic model was proposed in as early as 1956 by M. King Hubbert as a general model of the depletion of exhaustible resources, such as oil (Deffeyes 2008). The idea is that for a particular reserve of a resource, such as an oil field, the cumulative amount of the resource extracted through time follows a pattern resembling the logistic function. In the function (as depicted in Figure 10.6), n might be the total amount of oil extracted from an oil field by time t with m being the initial production amount and k being the total recoverable oil in the oil field. The amount extracted rises swiftly at first, because facilities for extraction (drills and pumps) are added to the location. But later, the rate of production slows as the amount of recoverable resource becomes depleted.

Let us think carefully about what is meant by the rate of production. Suppose the cumulative amount of oil extracted from an oil field by time t is the logistic function given by

$$n(t) = \frac{k}{1 + \left(\dfrac{k-m}{m}\right)e^{-bt}}.$$

The notation $n(t)$ here does not mean n times t but n is a function of t. Now suppose we wait a small amount of time s (perhaps a month, if time is measured in years) into the future so that the time becomes $t + s$. The cumulative amount of oil then would be the function evaluated at time $t + s$:

$$n(t+s) = \frac{k}{1 + \left(\dfrac{k-m}{m}\right)e^{-b(t+s)}}.$$

The amount of oil produced during the small amount of time s is the difference between $n(t + s)$ and $n(t)$. A rate is amount divided by time. The rate of oil production during the time interval of length s is the difference between $n(t + s)$ and $n(t)$ divided by s:

$$\frac{n(t+s)-n(t)}{s} = \left[\frac{k}{1 + \left(\dfrac{k-m}{m}\right)e^{-b(t+s)}} - \frac{k}{1 + \left(\dfrac{k-m}{m}\right)e^{-bt}}\right]/s.$$

This is a formula for calculating the rate of oil production during a small amount of time s beginning at time t. This rate will change over time, showing how the production rate changes over time.

Let us draw a graph of the rate of oil production versus time. The formula looks a bit complicated, but it should be easy to calculate in R if we are careful. Start the following script:

```
#================================================================
# R script to plot Hubbert's model of oil production.
#================================================================
k=100        # Maximum amount of recoverable resource.
b=.5         # Resource depleted faster if b is larger.
m=1          # Initial amount produced is m.
t=(0:100)*20/100   # Range of values of time t from 0 to 20.
s=.01        # Small interval of time.
change.n=k/(1+((k-m)/m)*exp(-b*(t+s)))-k/(1+((k-m)/m)*exp(-b*t))
             # Amount of oil extracted between time t and time
             # t+s.
rate.n=change.n/s  # Rate of oil production between time t and
                   # time t+s.
plot(t,rate.n,type="l",lty=1,xlab="time",ylab="rate of oil
  production")
```

When you run the script, Figure 10.7 results. We can see that the rate of oil production increases at first, then peaks, and finally declines as the resource becomes depleted. The curve in Figure 10.7 is often called "Hubbert's curve." In the section "Computational and Algebraic Challenges," you get the opportunity to overlay Hubbert's curve on some oil production data. In real life, you will have the opportunity to experience Hubbert's curve first hand.

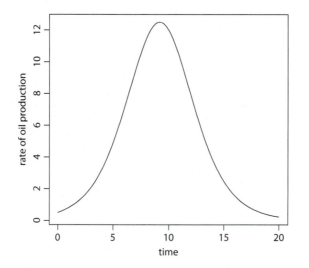

FIGURE 10.7
Graph of Hubbert's model of the rate of extraction of an exhaustible resource such as oil from an oil field.

Final Remarks

Biological quantities tend to grow multiplicatively, and so one encounters exponential and logarithmic functions throughout the life sciences. In statistics, probabilities are frequently multiplied, and statistical analyses are filled with exponential and logarithmic functions. Other sciences use exponential and logarithmic functions heavily as well. In the opinion of the author, these functions are often not given adequate time in high school and early undergraduate mathematics courses. Students' struggles with quantitative aspects of sciences can many times be traced to lack of experience with exponential and logarithmic functions. Get good with those functions, and your understanding of the natural world will increase exponentially.

WHAT WE LEARNED

1. The power function given by $y = x^a$, where x is any positive real number, is defined in mathematics for all real values of a. The power operator ^ in R accepts decimal numbers for powers:

```
> x=5
> a=1/(1:10)
> a
 [1] 1.0000000 0.5000000 0.3333333 0.2500000 0.2000000
 [6] 0.1666667 0.1428571 0.1250000 0.1111111 0.1000000
> y=x^a
> y
 [1] 5.000000 2.236068 1.709976 1.495349 1.379730
 [6] 1.307660 1.258499 1.222845 1.195813 1.174619
```

2. Algebraic properties of real-valued powers are the same as those of integer or rational powers:

$$x^0 = 1,$$
$$x^1 = x,$$
$$x^{-u} = 1/x^u,$$
$$x^u x^v = x^{u+v},$$
$$(x^u)^v = x^{uv}.$$

Here, x is a positive real number and u and v are any real numbers.

3. A special irrational number called e arises from the function $y = \left(1 + \dfrac{1}{x}\right)^x$ as the value of x increases toward positive infinity or decreases toward negative infinity. The value of e is approximately 2.718282.

4. The exponential function is $y = e^x$, where x is any real number. Sometimes the exponential function is denoted $y = \exp(x)$, especially when x is a complicated formula that would be hard to see when printed in a tiny exponent font. In R, the exponential function is exp() and, like the trigonometric functions, it is fully vectorized:

```
> exp(1)
[1] 2.718282
> x=-5:5
> exp(x)
 [1] 6.737947e-03 1.831564e-02 4.978707e-02
 [4] 1.353353e-01 3.678794e-01 1.000000e+00
 [7] 2.718282e+00 7.389056e+00 2.008554e+01
[10] 5.459815e+01 1.484132e+02
```

5. A frequently used mathematical model containing the exponential function is the exponential growth model given by

$$n = me^{rt},$$

where m is the initial amount of a quantity (bacteria, dollars, radioactive atoms), r is the rate of increase or decrease, and n is the amount of the quantity present after time t.

6. The natural logarithm of a positive real-valued quantity y is defined as $y = e^{\log(y)}$ or, in other words, the real number power to which e must be raised to get y. The logarithmic function is the exponential function $y = e^x$ in reverse, in which x is regarded as a function of y. Algebraically, the exponential and logarithmic functions undo each other: $\log(e^x) = x$, $e^{\log(y)} = y$. In R, the natural logarithm function is log():

```
> u=1:5
> v=log(u)
> v
[1] 0.0000000 0.6931472 1.0986123 1.3862944 1.6094379
> w=exp(v)
> w
[1] 1 2 3 4 5
```

7. The logarithm of a product of two positive quantities y and z is the sum of the logarithms of y and z: $\log(yz) = \log(y) + \log(z)$. The logarithm of a quotient of two positive quantities y and z is the difference between the logarithms of y and z: $\log(y/z) = \log(y) - \log(z)$:

```
> y=153
> z=227
> y*z
```

```
[1] 34731
> log(y*z)
[1] 10.45539
> log(y)+log(z)
[1] 10.45539
> exp(log(y)+log(z))
[1] 34731
```

8. Other positive numbers besides e can serve as the base for systems of logarithms, through the definition $y = a^{\log_a(y)}$. Along with base e logarithms, base 10 and base 2 logarithms are frequently seen in scientific applications. Base 10 and base 2 logarithm functions are available in R as `log10()` and `log2()`, respectively. Logarithms for any arbitrary base a can be calculated from natural logarithms as $\log_a(y) = \left[\log(y)\right] / \left[\log(a)\right]$. A function in R for $\log_a(y)$ is `log(y,a)`:

```
> log10(10000)
[1] 4
> log(10000)/log(10)
[1] 4
> log(10000,10)
[1] 4
```

9. In science, logarithmic scales are used for conveniently representing quantities that vary across tremendous ranges. The quantities in such scales are logarithms of the original quantities, and integer tic marks represent orders of magnitude. Different bases are used in logarithmic scales by tradition. Well-known logarithmic scales include pH in chemistry, Richter earthquake magnitude in geology, and star magnitude in astronomy.

Computational and Algebraic Challenges

10.1. Algebraically solve for x:

$$y = e^x \quad y = e^{-x} \quad y = e^{3-2x} \quad y = 15e^{3-2x} \quad y = ce^{a+bx} \quad y = 10 + ce^{a+bx} \quad y = \frac{1}{1+e^{-x}}$$

$$y = \log(x) \quad y = 5\log(x) \quad y = 6 + 8\log(x) \quad y = a + b\log(x) \quad y = 7\log\left(\frac{3+x}{4}\right) \quad y = 5^x.$$

10.2. a. For each of the following concentrations (moles/liter) of hydrogen ions, calculate the pH:

2.31e-10 5.43e-08 4.77e-06 8.19e-04 6.06e-03

b. For each of the following values of pH, calculate the concentration of hydrogen ions:

1.45 3.62 5.98 7.03 9.50 11.27

c. Calculate and draw a pair of graphs: pH as a function of hydrogen ion concentration, and hydrogen ion concentration as a function of pH. For each graph calculate enough values to produce smooth function curves.

10.3. Perform star magnitude calculations: How many times brighter than Sirius is Venus at its maximum apparent brightness? How many times brighter than Vega is Sirius? How many times brighter than Polaris is Sirius?

10.4. Perform radionuclide dating: Write an R function to calculate the age of a sample of a radioactive element. The function arguments should be half-life, initial amount, and present amount.

10.5. A population is growing according to the exponential growth equation. Algebraically determine the amount of time needed for the population to double in size. Use R and your formula to calculate the doubling time for a population with a growth rate of 0.04, compounded continuously. Draw a graph of such a population's size through time, starting at an initial population size of 100; use a long enough time interval for the population to at least double in size.

10.6. Newton's law of cooling is given by the following equation:

$$T = a + (T_0 - a)e^{-kt}.$$

Here, T is the temperature of an object at time t, T_0 is the initial temperature of the object, a is the ambient temperature, and k is a cooling rate constant with a value that depends on the properties of the object. A furnace in a house suddenly ceases to function. The temperature at the moment the heating stops is 74°F. The temperature outside is 25°F. The cooling rate constant for the house is $k = 0.1248$, and time is measured in hours. Draw a graph of the house's temperature for the next 36 hours.

10.7. Draw a plot of the U.S. population (census numbers appear in Chapter 1, Computational Challenge 1.3) superimposed with a logistic population growth curve. Use $t = 0, 1, 2, \ldots$ for the time scale, where 0 is 1780, 1 is 1790, 2 is 1800, and so on. Thus the first census corresponds with $t = 1$. Divide the population sizes by 1,000,000 so that the sizes represent millions of people. Use the following values for the constants in the logistic curve: $k = 432.7$, $m = 5.552$, and $b = 0.2275$.

10.8. In a famous set of experiments, the Russian biologist Gause (1934) tested the logistic population growth model in the laboratory. Some of his experimental data are given in this computational challenge. He grew populations of two species of *Paramecia* in growth media. There were three replicate cultures of each species. Each culture was inoculated with 2 cells per 0.5 cc of culture media. Starting from day 2, daily samples of 0.5 cc were drawn from the cultures and cells in the samples were counted. Columns 2 through 4 are *Paramecium aurelia* (cells per 0.5 cc) and columns 5 through 7 are *P. caudatum* (cells per 0.5 cc) cells. Plot the three populations of *P. aurelia* data on the same graph with a logistic curve superimposed. Use the following values for the constants in the logistic curve: $k = 559.3$, $b = 0.7812$, and $m = 2$. Plot the three populations of *P. caudatum* data on another graph with a logistic curve superimposed. Use the following values for the constants in the logistic curve: $k = 202.6$, $b = 0.6279$, and $m = 2$.

day	Pa1	Pa2	Pa3	Pc1	Pc2	Pc3
0	2	2	2	2	2	2
2	17	15	11	8	13	6
3	29	36	37	9	6	7
4	39	62	67	14	7	6
5	63	84	134	16	16	22
6	185	156	226	57	40	32
7	258	234	306	94	98	84
8	267	348	376	142	124	100
9	392	370	485	175	145	138
10	510	480	530	150	175	189
11	570	520	650	200	280	170
12	650	575	605	172	240	204
13	560	400	580	189	230	210
14	575	545	660	234	171	140
15	650	560	460	192	219	165
16	550	480	650	168	216	152
17	480	510	575	240	195	135
18	520	650	525	183	216	219
19	500	500	550	219	189	219

10.9. Annual oil production figures for Norway's oil fields in the North Sea (http://www.npd.no/en/Publications/Resource-Reports/2009/) are given. Draw a plot of these data over time with a Hubbert's oil production curve overlaid. Use $t = 0, 1, 2, \ldots$ for the time scale, where 0 is 1970, 1 is 1971, 2 is 1972, and so on. Use the following values for the constants in the Hubbert's curve: $m = 26.39$, $k = 4211$, and $b = 0.1727$. Also, the time increment s is 1.

```
year        1971 1972 1973 1974 1975 1976 1977 1978 1979 1980
production   0.4  1.9  1.9  2.0 11.0 16.2 16.6 20.6 22.5 28.2

1981 1982 1983 1984 1985 1986 1987 1988 1989 1990  1991  1992
27.5 28.5 35.6 41.1 44.8 48.8 57.0 64.7 86.0 94.5 108.5 124.0

 1993  1994  1995  1996  1997  1998  1999  2000  2001  2002
131.8 146.3 156.8 175.4 175.9 168.7 168.7 181.2 180.9 173.6

 2003  2004  2005  2006  2007  2008
165.5 162.8 148.1 136.6 128.3 122.7
```

References

Deffeyes, K. S. 2008. *Hubbert's Peak: The Impending World Oil Shortage*. Princeton: Princeton University Press.

Gause, G. F. 1934. *The Struggle for Existence*. Baltimore: Williams & Wilkins.

Macdougall, D. 2008. *Nature's Clocks: How Scientists Measure the Age of Almost Everything*. Berkeley: University of California Press.

Verhulst, P.-F. 1838. Notice sur la loi que la population poursuit dans son accroissement. *Correspondance Mathématique et Physique* 10: 113–121.

11

Matrix Arithmetic

Another Way to Multiply Vectors

We have seen that if x and y are vectors of numbers in R, then R defines the product x*y as the vector containing the elements of x each multiplied by the corresponding elements of y. This operation is called **elementwise** multiplication.

There is another kind of multiplication for vectors called the dot product or the scalar product that comes in handy in many applications in science and commerce. Here is an example to introduce the dot product. You will recall that in Chapter 6 (Loops), we studied the calculations for projecting the sizes of age-structured wildlife populations into the future. Suppose a population of a small mammal species is represented by the vector $\mathbf{n} = (n_1, n_2, n_3, n_4)$, containing, respectively, the numbers of 0-year-olds, 1-year-olds, 2-year-olds, and animals 3 years old or older. Suppose the vector $\mathbf{f} = (f_1, f_2, f_3, f_4)$ contains the average number of offspring born in the population per year to an average 0-year-old, 1-year-old, 2-year-old, and 3-or-more-year-old, respectively (probably the first two birth rates in the vector are zero or very low because they are birth rates of young population members). We will use boldface letters like $\mathbf{n}$ and $\mathbf{f}$ to denote vectors in mathematical formulas, as opposed to the font used for implementation of the vectors and formulas in R statements. Recall that the average total number of offspring born in the population in a year would be calculated as follows:

$$\text{total offspring} = n_1 f_1 + n_2 f_2 + n_3 f_3 + n_4 f_4.$$

If the numerical values were available for $\mathbf{n}$ and $\mathbf{f}$, the calculation could be accomplished in R with the following statements:

```
> n=c(49,36,28,22)
> f=c(0,.2,2.3,3.8)
> total.offspring=sum(n*f)
> total.offspring
[1] 155.2
```

The result of this calculation is a single number, or a **scalar**, in the jargon of matrix and vector arithmetic. The calculation represented by sum(n*f) is the

dot (or **scalar**) **product** of two vectors n and f of identical lengths. In general, if $\mathbf{x} = (x_1, x_2, ..., x_k)$ and $\mathbf{y} = (y_1, y_2,..., y_k)$ are vectors each with length k, the symbol for a dot product is a centered dot, and the general definition of the dot product is the scalar number resulting from the following formula:

$$\mathbf{x} \bullet \mathbf{y} = x_1 y_1 + x_2 y_2 + \cdots + x_k y_k.$$

R has a special operator, "%*%", that returns the dot product of two vectors:

```
> n%*%f
       [,1]
[1,] 155.2
```

The forthcoming matrix calculations in this chapter are going to produce arrays of answers in rows and columns, and the output of the above operation states that row 1 and column 1 of the answer is 155.2.

The dot product calculation has many uses. If the vector **x** contains the numbers of different models of automobiles produced by a manufacturer and the vector **y** contains the profit per automobile for each model, then $\mathbf{x} \circ \mathbf{y}$ gives the total profit resulting from the mix of auto models produced. If the vector **x** contains quantities that are recorded about a house, such as the number of square feet, the number of bedrooms, the number of bathrooms, the distance to the nearest high school, and so on, and the vector **b** contains the average contribution to the sale value of the house by a unit of each quantity in **x**, then $\mathbf{b} \circ \mathbf{x}$ is the estimated sale value (appraised value) of the house. A remarkable result from geometry, which we will not prove here, is that if **r** contains coordinates of a point in two dimensions, three dimensions, or any number of higher dimensions, and **s** contains the coordinates of another such point in the same coordinate system, and θ is the angle between the two line segments drawn from the origin to the points represented by **r** and **s**, then $\mathbf{r} \circ \mathbf{s} / \left[\left(\sqrt{\mathbf{r} \circ \mathbf{r}} \right) \left(\sqrt{\mathbf{s} \circ \mathbf{s}} \right) \right]$ is the cosine of θ:

```
> cos(pi/4)
[1] 0.7071068
> r=c(2,2)    # Line segments from (0,0) to r and to s
> s=c(2,0)    #   form an angle of pi/4 (45 degrees).
> r%*%s/(sqrt(r%*%r)*sqrt(s%*%s))
          [,1]
[1,] 0.7071068
```

There is frequent occasion to perform and keep track of many dot products. For instance, denote by p_1 the average yearly survival probability for 0-year-olds in the mammal population described earlier, and let p_2, p_3, and p_4 be the respective annual survival probabilities of 1-year-olds, 2-year-olds, and animals 3 years old or older. Define the vectors $\mathbf{p}_1 = (p_1, 0, 0, 0)$, $\mathbf{p}_2 = (0, p_2, 0, 0)$, and $p_3 = (0, 0, p_3, p_4)$. Then after a year has elapsed, $p_1 \smile n$ $(= p_1 n_1)$ is the number of 1-year-olds, $p_2 \smile n$ is the number of 2-year-olds, and $p_3 \smile n$ is the number of animals 3 years old or older. As well, $n \circ f$ is the number of 0-year olds after a year has elapsed. Wildlife scientists compactly represent the calculations that project an age-structured population forward 1 year in time by stacking the vectors f, p_1, p_2, and p_3 as rows of numbers into a matrix. The four dot products $n \circ f$, $p_1 \smile n$, $p_2 \smile n$, and $p_3 \smile n$ for projecting the population in 1 year become the components of an operation called matrix multiplication.

Matrix Multiplication

A **matrix** is a rectangular array of numbers. Matrices are simple, really, except that matrix multiplication is defined in a manner that seems unintuitive at first. We can understand matrix multiplication as a way of performing and keeping track of many dot products of vectors, a task which has proven enormously useful in quantitative science.

First, we note that R can handle not just vectors but whole matrices. We can build a matrix in R by binding vectors together into a matrix as rows or as columns using the rbind() and cbind() commands. Try them:

```
> x1=c(3,4,5,6)
> x2=c(10,11,12,13)
> x3=c(-1,-2,-3,-4)
> A=rbind(x1,x2,x3)
> B=cbind(x1,x2,x3)
> A
     [,1] [,2] [,3] [,4]
x1      3    4    5    6
x2     10   11   12   13
x3     -1   -2   -3   -4
> B
     x1  x2  x3
[1,]  3  10  -1
[2,]  4  11  -2
[3,]  5  12  -3
[4,]  6  13  -4
```

The matrix **A** defined in the above R statements has three rows and four columns (we say that **A** is a 3×4 matrix), whereas the matrix **B** is 4×3. You can see that when **A** and **B** were printed to the console, the rows of **A** and the columns of **B** were labeled with the original vector names, for the convenience of recognizing where the original vectors are located. However, matrix elements are generally referenced by their row and column numbers. In R, individual elements can be picked out of a matrix using their row and column numbers in square brackets, with the row number always appearing first, just like it is done for data frames:

```
> A[2,3]
x2
12
> B[4,3]
x3
-4
> A[2,3]+B[4,3]
x2
 8
```

Like the provision for extracting whole portions of data from data frames, submatrices can be extracted from matrices by referencing vectors of row and column numbers, and all rows or columns are referenced by omitting the row or column number entirely:

```
> A[1,2:4]
[1] 4 5 6
> B[c(1,3),1:3]
     x1 x2 x3
[1,]  3 10 -1
[2,]  5 12 -3
> A[2,]
[1] 10 11 12 13
```

In R, a matrix differs from a data frame in that a matrix can only contain numerical elements, while a data frame can have categorical or numerical data.

The **matrix product AB** of a matrix **A** ($l \times m$) and a matrix **B** ($m \times n$) is defined if the number of columns of **A** equals the number of rows of **B**. Think of the first matrix **A** in the product as a stack of vectors, each with m elements, in the form of rows:

$$\mathbf{A} = \begin{bmatrix} \mathbf{a}_1 \\ \mathbf{a}_2 \\ \vdots \\ \mathbf{a}_l \end{bmatrix}.$$

Think of the second matrix in the product as a line of vectors $\mathbf{b}_1, \mathbf{b}_2, \ldots, \mathbf{b}_n$, each with m elements, in the form of columns:

$$\mathbf{B} = \begin{bmatrix} \mathbf{b}_1 & \mathbf{b}_2 & \cdots & \mathbf{b}_n \end{bmatrix}.$$

The matrix product **AB** is a matrix with l rows and n columns, consisting of all the pairwise dot products of the vectors making up **A** and **B**:

$$AB = \begin{bmatrix} a_1 \cdot b_1 & a_1 \cdot b_2 & \cdots & a_1 \cdot b_n \\ a_2 \cdot b_1 & a_2 \cdot b_2 & \cdots & a_2 \cdot b_n \\ \vdots & \vdots & \ddots & \vdots \\ a_l \cdot b_1 & a_l \cdot b_2 & \cdots & a_l \cdot b_n \end{bmatrix}$$

In other words, the element in the ith row and jth column of **AB** is the dot product of the ith row of **A** and the jth column of **B**.

Matrix multiplication is a laborious computational task, although you should try multiplying a few small matrices by hand (perhaps aided by a calculator) just to get a feel for the concept. For instance, take the matrices **A** and **B** from the previous R statements and find the product by hand. Then, check your calculations using R. In R, the operator %*% that we used for dot product also performs matrix multiplication:

```
> C=A%*%B
> C
     x1    x2    x3
x1   86   212   -50
x2  212   534  -120
x3  -50  -120    30
```

In general, matrix multiplication is not commutative, that is, **AB** $\neq$ **BA**. The product **BA** could be a different-sized matrix or could even be undefined if the rows of **B** and columns of **A** do not match. For the matrices **A** and **B** in the above R statements, the product **BA** is a 4×4 matrix:

```
> D=B%*%A
> D
      [,1]  [,2]  [,3]  [,4]
[1,]   110   124   138   152
[2,]   124   141   158   175
[3,]   138   158   178   198
[4,]   152   175   198   221
```

When discussing a matrix product **AB**, saying that **A** is "postmultiplied" by **B** and that **B** is "premultiplied" by **A** helps to avoid confusion.

Although matrix multiplication is not commutative, it is associative. You can put parentheses anywhere in a string of matrix products:

$$ABC = (AB)C = A(BC).$$

A matrix with only one row or one column is called, respectively, a **row vector** or a **column vector** in matrix terminology. When a matrix is postmultiplied by a column vector, the result is a column vector. When a matrix is

premultiplied by a row vector, the result is a row vector. In R, an ordinary
vector has no row or column distinction. However, in R if you postmultiply
a matrix with a vector, R will treat the vector as a column vector, and if you
premultiply a matrix with a vector, R will treat the vector as a row vector:

```
> b=c(2,0,4,-1)
> A%*%b
    [,1]
x1    20
x2    55
x3   -10
> c=c(2,-1,4)
> c%*%A
      [,1] [,2] [,3] [,4]
[1,]    -8  -11  -14  -17
```

Also, if you pick out a row or a column of a matrix in R, then R treats the
result as just an ordinary vector without any row or column distinction:

```
> A[1,]
[1] 3 4 5 6
> B[,1]
[1] 3 4 5 6
```

One more multiplication type involving matrices must be noted now. The
scalar multiplication of a matrix is defined as a matrix, say **A**, multiplied
by a scalar number, say x; the operation is denoted **A**x or x**A** and results in
a matrix containing all the elements of **A**, each individually multiplied by
x. In R, scalar multiplication of a matrix is accomplished with the ordinary
multiplication operator *:

```
> x=2
> x*A
      [,1] [,2] [,3] [,4]
x1      6    8   10   12
x2     20   22   24   26
x3     -2   -4   -6   -8
```

Scalar multiplication of a matrix is commutative: **A**$x = x$**A**.

Matrix Addition and Subtraction

Unlike matrix multiplication, **matrix addition** is defined elementwise, as
we intuitively suppose it should be. In matrix addition, each element of one
matrix is added to the corresponding element of another matrix. Matrix
addition is defined only for two matrices with the same number of rows and

the same number of columns. **Matrix subtraction** is similar to matrix addition, involving subtraction of elements of one matrix from the corresponding elements of another matrix. The ordinary plus and minus signs in R work for matrix addition and subtraction:

```
> k=1:10
> A=matrix(k,2,5)
> A
     [,1] [,2] [,3] [,4] [,5]
[1,]    1    3    5    7    9
[2,]    2    4    6    8   10
> j=c(1,2)
> B=matrix(j,2,5)
> B
     [,1] [,2] [,3] [,4] [,5]
[1,]    1    1    1    1    1
[2,]    2    2    2    2    2
> A+B
     [,1] [,2] [,3] [,4] [,5]
[1,]    2    4    6    8   10
[2,]    4    6    8   10   12
> A-B
     [,1] [,2] [,3] [,4] [,5]
[1,]    0    2    4    6    8
[2,]    0    2    4    6    8
```

The above statements made use of the `matrix()` function in R for building matrices. The statement A=matrix(k,2,5) reads the vector k into a 2×5 matrix named A, column by column. The statement B=matrix(j,2,5) reads the vector j into a 2×5 matrix named B, column by column, and the statement shows that if the vector being read in is too small, it will be recycled until the matrix is filled up.

Unlike matrix multiplication, matrix addition is commutative: $\mathbf{A} + \mathbf{B} = \mathbf{B} + \mathbf{A}$. Subtraction is more properly thought of as addition and in that proper sense is commutative: $\mathbf{A} + (-\mathbf{B}) = (-\mathbf{B}) + \mathbf{A}$. Here, $-\mathbf{B}$ is interpreted as the scalar multiplication of $\mathbf{B}$ by -1.

Reading a Data File into a Matrix

A file of data can be read into a matrix with the `matrix()` function. Suppose the file `data.txt` is a space-separated text file in the working directory of R. The file should have only numerical data. The statement

```
A=matrix(scan("data.txt"),nrow=6,ncol=8,byrow=TRUE)
```

builds A as a matrix with six rows and eight columns. The argument `scan()` reads the data file row by row, and so an additional argument `byrow=TRUE` is needed if the data are to be entered into the matrix row by row instead of the default method of column by column. The number of rows and columns specified in the `matrix()` function should match those of the data file if the matrix is supposed to look like the file.

Real-World Example

The age-structured wildlife population example from Chapter 6 (Loops) is ready-made to be reformulated in terms of matrices. You will recall that the wildlife population in Chapter 6 had three age classes: juveniles (less than 1 year old), subadults (nonbreeding animals between 1 and 2 years old), and breeding adults (2 years old and older). The numbers of juveniles, subadults, and adults in the population at time t were denoted respectively by J_t, S_t, A_t. These age classes were projected one time unit (year) into the future with three equations:

$$J_{t+1} = fA_t,$$

$$S_{t+1} = p_1 J_t,$$

$$A_{t+1} = p_2 S_t + p_3 A_t.$$

Here, p_1, p_2, and p_3 are the annual survival probabilities for individuals in the three age classes, and f is the average annual number of offspring produced by each adult (fecundity). We can rewrite these equations to make them look like dot products:

$$J_{t+1} = 0J_t + 0S_t + fA_t \quad \text{(dot product of } (0, 0, f) \text{ and } (J_t, S_t, A_t)),$$

$$S_{t+1} = p_1 J_t + 0S_t + 0A_t \quad \text{(dot product of } (p_1, 0, 0) \text{ and } (J_t, S_t, A_t)),$$

$$A_{t+1} = 0J_t + p_2 S_t + p_3 A_t \quad \text{(dot product of } (0, p_2, p_3) \text{ and } (J_t, S_t, A_t)).$$

You are correct if you smell a matrix multiplication lurking somewhere! Gather the vector (J_t, S_t, A_t) of age classes into a column vector (let us call it $\mathbf{n}_t$):

$$\mathbf{n}_t = \begin{bmatrix} J_t \\ S_t \\ A_t \end{bmatrix}.$$

Gather the survival probabilities and fecundity rates into a matrix (let us call it **M**):

$$\mathbf{M} = \begin{bmatrix} 0 & 0 & f \\ p_1 & 0 & 0 \\ 0 & p_2 & p_3 \end{bmatrix}.$$

The column vector $\mathbf{n}_{t+1}$ of next year's age classes is found to be a matrix multiplication:

$$\mathbf{n}_{t+1} = \mathbf{M}\mathbf{n}_t.$$

Isn't that elegant? The three projection equations are expressed compactly in matrix form. Take a piece of paper and write out the matrix multiplication on the right-hand side of the equation to see how it corresponds to the three projection equations, if you do not see it already.

Let us rewrite the script from Chapter 6 that calculated and plotted the age classes of the Northern Spotted Owl through time, using the matrix capabilities of R. We will represent the projection calculations in matrix form. The exercise is an opportunity to introduce the matplot() function, which plots every column of one matrix versus the corresponding column of another matrix. The matplot() function otherwise resembles and accepts all the graphical arguments of the plot() statement.

Here is a rewritten script for the population projection using matrix calculations:

```
#==============================================================
# R script to calculate and plot age class sizes through
# time for an age-structured wildlife population, using
# the matrix projection model. Demographic rates for the
# Northern Spotted Owl are from Noon and Biles
# (Journal of Wildlife Management, 1990).
#==============================================================
num.times=20
num.ages=3

p1=.11   # Age-specific survival and fecundity rates.
p2=.71   #
p3=.94   #
f=.24    #

M=rbind(c(0,0,f),c(p1,0,0),c(0,p2,p3)) # Projection matrix.

N=matrix(0,num.times,num.ages) # num.times X num.ages matrix of 0's.
                               #  Rows of N will be the population
                               #  vectors through time.
```

```
N[1,]=c(1200,800,2000) # First row of N contains initial sizes.


for (t in 1:(num.times-1)) {
  N[t+1,]=M%*%N[t,]
}
N   # Print N to the console.


time.t=0:(num.times-1)
#------------------------------------------------------------------
# Following matplot() function plots every column of N (vertical
# axis) versus time.t (horizontal axis) & accepts all graphical
# arguments of the plot() function.
#------------------------------------------------------------------
matplot(time.t,N,type="l",lty=c(2,5,1),
    xlab="time in years",ylab="population size",ylim=c(0,2600))
```

In the script, N is set up as a matrix of zeros with 20 (num.times) rows and 3 (num.ages) columns. The first row of N is then assigned to be the initial sizes of the age classes. In the for loop, each row of N is projected with a matrix multiplication, with the projection results filling the next row below. Note that in the matrix multiplication in the statement

```
N[t+1,]=M%*%N[t,],
```

N[t,] is regarded by R as a column vector even though it was referenced as a row of N. This is because when a row or a column of a matrix, like the row N[t,], is referenced in R, the result is an ordinary vector with no row or column property. Then the multiplication M%*%N[t,] interprets the ordinary vector N[t,] as a column vector.

The matplot() function works like the plot() function, except that the matplot() function accepts matrices in the role of the horizontal and vertical axis variables. Here, a vector (horizontal axis) is plotted against each column of a matrix (vertical axis), all on one graph with one set of axes. Each resulting pair of vectors is plotted in a different color. The symbols and graphical parameters can be altered with the regular arguments as used in the plot() command. Some of the graphical settings accept vector arguments. An example appears in the line type setting lty=c(2,5,1), corresponding respectively to dashed, long dashed, and solid line types. The three line plots on the graph will use in turn the line types designated, in order of appearance. Some experimentation with the limits of the vertical axis (ylim=c(0,2600)) is usually necessary to get a graph that displays all the individual plots well.

The resulting graph (Figure 11.1) is mostly identical to Figure 6.1. The version of Figure 11.1 arising from matplot() that appears on your computer,

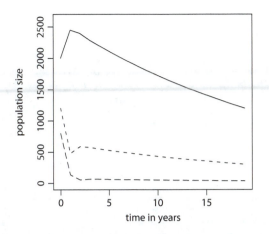

FIGURE 11.1
Female adult (solid line), female subadult (long dashes), and female juvenile abundances (short dashes) of the Northern Spotted Owl (*Strix occidentalis caurina*) projected 20 years with the matrix projection model using survival and fecundity data. (From Noon, B. R. and Biles, C. M., *Journal of Wildlife Management*, 54, 18–27, 1990.)

however, will show the lines in different colors. Insert the optional argument `col="black"` in `matplot()` to suppress the colors, if you wish.

Final Remarks

Getting familiar with matrix multiplication and other matrix properties will take you a long, long way in science. A substantial portion of statistics and mathematical modeling is formulated in terms of matrices. In Chapter 12, we will get to know one of the matrix "killer apps": solving systems of linear equations.

> The Matrix is everywhere. It is all around us. Even now, in this very room.
>
> **—Morpheus, in *The Matrix* (Wachowski and Wachowski, 1999).**

WHAT WE LEARNED

1. If $\mathbf{x} = (x_1, x_2, \ldots, x_k)$ and $\mathbf{y} = (y_1, y_2, \ldots, y_k)$ are vectors each with length k, the definition of the dot product is the scalar number resulting from the following formula:

$$\mathbf{x} \cdot \mathbf{y} = x_1 y_1 + x_2 y_2 + \cdots + x_k y_k.$$

In R, the dot product of vectors x and y can be obtained by sum(x*y) or by x%*%y.

2. The **matrix product AB** of a matrix **A** ($l \times m$) and a matrix **B** ($m \times n$) is defined if the number of columns of **A** equals the number of rows of **B**. The matrix product **AB** is a matrix with l rows and n columns, consisting of all the pairwise dot products of the vectors making up the rows of **A** and the columns of **B**:

$$\mathbf{AB} = \begin{bmatrix} \mathbf{a}_1 \cdot \mathbf{b}_1 & \mathbf{a}_1 \cdot \mathbf{b}_2 & \cdots & \mathbf{a}_1 \cdot \mathbf{b}_n \\ \mathbf{a}_2 \cdot \mathbf{b}_1 & \mathbf{a}_2 \cdot \mathbf{b}_2 & \cdots & \mathbf{a}_2 \cdot \mathbf{b}_n \\ \vdots & \vdots & \ddots & \vdots \\ \mathbf{a}_l \cdot \mathbf{b} & \mathbf{a}_l \cdot \mathbf{b}_2 & \cdots & \mathbf{a}_l \cdot \mathbf{b}_n \end{bmatrix}$$

In other words, the element in the ith row and jth column of **AB** is the dot product of the ith row of **A** and the jth row of **B**. Such matrix multiplication is not in general commutative: **AB** ≠ **BA**. In R, the matrices **A** and **B** are multiplied with A%*%B.

3. The multiplication of a matrix **A** by a scalar x is denoted **A**x or x**A** and results in a matrix containing all the elements of **A**, each individually multiplied by x. In R, scalar multiplication of a matrix is accomplished with the ordinary multiplication operator *: x*A.

4. Matrix addition is defined elementwise. In matrix addition, each element of one matrix is added to the corresponding element of another matrix. Matrix addition is defined only for two matrices with the same number of rows and the same number of columns. Matrix subtraction is similar to matrix addition, involving subtraction of elements of one matrix from the corresponding elements of another matrix. The ordinary plus and minus signs in R work for matrix addition and subtraction: A+B, A-B.

5. A file of data can be read into a matrix with the matrix() function. Suppose the file data.txt is a space-separated text file in the working directory of R. The file should have only numerical data. The statement

```
A=matrix(scan("data.txt"),nrow=6,ncol=8,byrow=TRUE)
```

builds A as a matrix with six rows and eight columns.

Computational Challenges

11.1. Find the dot products of the following pairs of vectors:

$$\mathbf{x} = (3,12,7,-4,-9), \quad \mathbf{y} = (-2,0,4,8,-3).$$

$$\mathbf{x} = (0.3,0.5,0.2,0.4), \quad \mathbf{y} = (127,48,205,76).$$

$$\mathbf{x} = (1,1,1,1,1,1), \quad \mathbf{y} = (2,4,3,5,4,6).$$

11.2. Find the matrix product **AB** of the following matrices:

$$\mathbf{A} = \begin{bmatrix} 2 & -3 & 0 \\ 4 & 1 & -5 \end{bmatrix}, \quad \mathbf{B} = \begin{bmatrix} 7 & 6 \\ 2 & 4 \\ -8 & 1 \end{bmatrix}.$$

$$\mathbf{A} = \begin{bmatrix} 14 & 20 & 12 \\ 7 & 19 & 32 \\ 10 & 22 & 17 \end{bmatrix}, \quad \mathbf{B} = \begin{bmatrix} 0.23 & 0.32 \\ 0.14 & 0.19 \\ 0.04 & 0.22 \end{bmatrix}.$$

$$\mathbf{A} = \begin{bmatrix} 14 & 20 & 12 \\ 7 & 19 & 32 \\ 10 & 22 & 17 \end{bmatrix}, \quad \mathbf{B} = \begin{bmatrix} 1 & 0 & 0 \\ 0 & 1 & 0 \\ 0 & 0 & 1 \end{bmatrix}.$$

$$\mathbf{A} = \begin{bmatrix} 1 & 1 & 1 \\ 6 & 4 & 8 \\ 5 & 12 & 4 \\ 2 & 1 & 9 \end{bmatrix}, \quad \mathbf{B} = \begin{bmatrix} 1 & 6 & 5 & 2 \\ 1 & 4 & 12 & 1 \\ 1 & 8 & 4 & 9 \end{bmatrix}.$$

$$\mathbf{A} = \begin{bmatrix} 3 & 1 \\ 4 & 4 \end{bmatrix}, \quad \mathbf{B} = \begin{bmatrix} 1 & -0.5 \\ -2 & 1.5 \end{bmatrix}.$$

11.3. Redo Computational Challenge 6.2 (blue whale population projection) in matrix form. To formulate properly, start with the projection equations, then recast them as dot products of vectors, then put the vectors into matrices.

11.4. Redo Computational Challenge 6.3 (black spruce population projection) in matrix form.

Afternote

The theory and application of matrix projection models are a huge part of conservation biology and wildlife management (Caswell 2001). If you are interested in environmental science, there is a matrix in your future.

References

Caswell, H. 2001. *Matrix Population Models: Construction, Analysis, and Interpretation*, 2nd edition. Sunderland: Sinauer.

Noon, B. R., and C. M. Biles. 1990. Mathematical demography of Spotted Owls in the Pacific Northwest. *Journal of Wildlife Management* 54:18–27.

Wachowski, A. and L. Wachowski (Directors). 1999. *The Matrix*. Burbank, CA: Warner Bros. Pictures.

12

Systems of Linear Equations

Matrix Representation

Many high school algebra courses devote time to systems of linear equations. Let us look at the following example:

$$-x_1 + 4x_2 = 8,$$

$$3x_1 + 6x_2 = 30.$$

Here, x_1 and x_2 are unknown quantities. Some algebra books denote x_1 and x_2 as x and y (or with other symbols); we use subscripts instead in anticipation of casting the quantities as elements of a vector. Now, each equation is the equation for a line. In other words, the set of points with coordinates (x_1, x_2) on a two-dimensional Cartesian graph that satisfy the first equation is a line. Similarly, the set of points with coordinates x_1, x_2 on a two-dimensional Cartesian graph that satisfy the second equation is also a line, different from the first. We can rearrange the two equations a little to see better that the equations are in the more familiar form $x_2 = (intercept) + (slope) \times x_1$:

$$x_2 = 2 + \frac{1}{4}x_1,$$

$$x_2 = 5 - \frac{1}{2}x_1.$$

By a **solution to the system of linear equations**, we mean a set of values of x_1, x_2 that satisfy both equations simultaneously, that is, a set of points (x_1, x_2) that lie on both lines at the same time. Such a solution can happen two ways: (1) the lines intersect at a point or (2) the lines coincide. If, on the other hand, the lines are parallel, they will never meet and no solution will exist.

Let us draw a quick graph of these two lines. Start a small script:

```
x1=(0:100)*8/100   # Range of x1 values from 0 to 8.
x2.A=2+(1/4)*x1    # x2 values for line A.
x2.B=5-(1/2)*x1    # x2 values for line B
```

```
x2=cbind(x2.A,x2.B)   # Combine both sets of x2 values
                      #  into a matrix, as columns.
matplot(x1,x2,type="l",lty=c(1,1),col=c(1,1))
                      # Plot both columns of x2 vs x1,
                      #  line type (lty) solid (1),
                      #  color (col) black (1).
```

Run the script and view the result (Figure 12.1). The first line has a vertical axis intercept of 2 and an increasing slope, whereas the second line has an intercept of 5 and a decreasing slope. The lines intersect at a point, which appears to be at $x_1 = 4$ and $x_2 = 3$. It is easy to substitute these values into each of the equations to verify that the point $(4, 3)$ is in fact the solution to the system of two equations.

You might have seen already that the two equations can be expressed in terms of matrices. Collect the coefficients of x_1 and x_2 from the original two equations into a matrix, put x_1 and x_2 into a column vector, and put the constants 8 and 30 on the right-hand side of the equations into a column vector. The two equations can be expressed as the following matrix equation:

$$\begin{bmatrix} -1 & 4 \\ 3 & 6 \end{bmatrix}\begin{bmatrix} x_1 \\ x_2 \end{bmatrix}=\begin{bmatrix} 8 \\ 30 \end{bmatrix}.$$

Write out the matrix multiplication if you do not see yet how the matrix equation is actually our two simultaneous linear equations. If the coefficient matrix was denoted by **A**, the column vector of unknowns by **x**, and the column vector of constants by **c**, we can symbolically write the matrix equation as

$$\mathbf{Ax = c}.$$

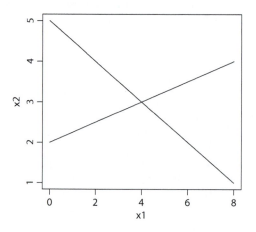

FIGURE 12.1

Plots of two linear equations: $-x_1 + 4x_2 = 8$ (positive slope) and $3x_1 + 6x_2 = 30$ (negative slope).

You will remember (or will discover, sometime soon in your studies) that the algebra for solving simultaneous linear equations can get tedious. One solves for the value of one variable algebraically or by elimination, then substitutes the value back into the system and solves again for another variable, and so on. Two unknowns is not hard, but three equations in three unknowns takes half a page of scratch paper, and four equations in four unknowns is hard to complete without making errors. Imagine, wouldn't it be nice if we could just divide both sides of the above matrix equation by **A**?

Well, the bad news is, there is no such thing as matrix division. But there is good news: we can multiply both sides of the matrix equation by the inverse of the matrix **A**.

Matrix Inverse

Something akin to matrix "division" for square matrices (matrices with same numbers of rows and columns) can be defined by analogy to ordinary division of real numbers. Ordinary division of real numbers is actually multiplication. For multiplication of real numbers, the **multiplicative identity** is the number 1, that is, any real number a multiplied by 1 is just a: $a(1) = a$. The **reciprocal**, or **inverse**, of the number a is another real number (call it b) such that if you multiply it by a you get the multiplicative identity: $ba = ab = 1$. We know this number b as $1/a$ or a^{-1}. Reciprocals do not exist for all real numbers; in particular, there is no reciprocal for 0. The key idea for extending division to matrices is that division by a real number a is multiplication by a^{-1}.

For matrix multiplication, a special square matrix called the **identity matrix** is the multiplicative identity. An identity matrix with k rows and k columns is universally denoted with the letter **I** and takes the form

$$\mathbf{I} = \begin{bmatrix} 1 & 0 & 0 & \cdots & 0 \\ 0 & 1 & 0 & \cdots & 0 \\ 0 & 0 & 1 & \cdots & 0 \\ \vdots & \vdots & \vdots & \ddots & \vdots \\ 0 & 0 & 0 & \cdots & 1 \end{bmatrix},$$

that is, a square matrix with 1s going down the upper left to lower right diagonal (frequently called the "main diagonal" in matrix lingo) and 0s everywhere else. If **C** is any matrix (square or otherwise), and an identity matrix of the right size is constructed so that the matrix multiplication is defined, then **CI** = **C** (where the number of rows of **I** equals the number of columns of **C**) as well as **IC** = **C** (where the number of columns of **I** equals the number of rows of **C**).

The inverse of a square matrix $\mathbf{A}$, denoted $\mathbf{A}^{-1}$, is a square matrix of the same size as $\mathbf{A}$ that produces an identity matrix when pre- or post-multiplied by $\mathbf{A}$:

$$\mathbf{A}\mathbf{A}^{-1} = \mathbf{A}^{-1}\mathbf{A} = \mathbf{I}.$$

For instance, here is the coefficient matrix from our system of equations above:

$$\mathbf{A} = \begin{bmatrix} -1 & 4 \\ 3 & 6 \end{bmatrix}.$$

Take a few seconds and verify by hand that the following matrix is the inverse of $\mathbf{A}$:

$$\mathbf{A}^{-1} = \begin{bmatrix} -\dfrac{6}{18} & \dfrac{4}{18} \\ \dfrac{3}{18} & \dfrac{1}{18} \end{bmatrix}.$$

If we know (or somehow can calculate) the inverse of $\mathbf{A}$, then that is the ticket for the solution to the system of linear equations. Take the matrix form of the system, $\mathbf{A}\mathbf{x} = \mathbf{c}$, and premultiply both sides of the equation by the inverse of $\mathbf{A}$:

$$\mathbf{A}^{-1}\mathbf{A}\mathbf{x} = \mathbf{A}^{-1}\mathbf{c}.$$

Because of the associative property of matrix multiplication, on the left-hand side we can do the $\mathbf{A}^{-1}\mathbf{A}$ multiplication first, giving

$$\mathbf{I}\mathbf{x} = \mathbf{A}^{-1}\mathbf{c}.$$

But $\mathbf{I}\mathbf{x}$ is just $\mathbf{x}$; we have solved the system! Written out,

$$\mathbf{x} = \mathbf{A}^{-1}\mathbf{c}$$

or

$$\begin{bmatrix} x_1 \\ x_2 \end{bmatrix} = \begin{bmatrix} -\dfrac{6}{18} & \dfrac{4}{18} \\ \dfrac{3}{18} & \dfrac{1}{18} \end{bmatrix} \begin{bmatrix} 8 \\ 30 \end{bmatrix} = \begin{bmatrix} 4 \\ 3 \end{bmatrix},$$

which is the solution we suspected from the graph (Figure 12.1). Do the last matrix multiplication by hand for practice.

Systems of linear equations can have more than two unknown quantities. A linear equation with three unknowns is the equation for a plane in three dimensions. Three equations in three unknowns can have a unique point solution; look at a corner of the room you are in where two adjacent walls and a ceiling meet. Four equations in four unknowns, or k equations in k unknowns, can have unique point solutions as well. Linear equations with four or more unknowns are called hyperplanes and cannot be envisioned well in our three-dimensional world. However, describing things with a vector of four or more numbers is pretty routine in our world; think of all the measurements you need to be fitted with a new outfit or new suit.

Whatever number of unknowns there are in the system of equations, the system has a matrix representation of the form $\mathbf{Ax} = \mathbf{c}$. We will concentrate on systems in which the number of unknowns is the same as the number of equations, that is, in which the matrix $\mathbf{A}$ of coefficients is square. Solving a system of k equations in k unknowns can be summarized in the form of the following mathematical result. Proving the result is beyond the scope of this book, but we will put the result to good use:

Result. The system of linear equations defined by $\mathbf{Ax} = \mathbf{c}$, where $\mathbf{A}$ is a $k \times k$ matrix and at least one of the elements of $\mathbf{c}$ is nonzero, has a unique point solution if an inverse matrix for $\mathbf{A}$ exists. The solution is then given by $\mathbf{x} = \mathbf{A}^{-1}\mathbf{c}$.

So, solving a system of linear equations boils down to finding an inverse matrix. Unfortunately, the world is not always just, and reliable matrix inversion turns out to be a challenging numerical problem that is still an active area of applied mathematics research (enormous matrices that are only sparsely populated with nonzero elements are a particular challenge). Fortunately, the most common problems of everyday science are routinely handled by our contemporary matrix inversion algorithms. Naturally, R contains a function for taking care of most of your matrix inversion needs.

Inverse Matrices and System Solutions in R

The solve() function in R can be used in two different ways: (1) to solve a system of linear equations, if a solution exists, or (2) to calculate the inverse of a square matrix, if the inverse exists. Both tasks are accomplished within the constraints of floating point arithmetic (which creates some round-off error), and either might fail in situations (such as large sparse matrices) in which round-off error tends to build up quickly.

The following small script illustrates how to use the solve() function for the two tasks, using our example system of two equations from Figure 12.1:

```
#===================================================================
# R script to: (a) Solve a system of k equations in k unknowns,
# represented by the matrix equation Ax = c, where A is a k X k
# matrix c is a k X 1 column vector of constants, and x is a k X 1
```

```
# column vector of unknown values. (b) Calculate the inverse of the
# matrix A, if an inverse exists.
#=================================================================

#------------------------------------------------------------------
#  Enter rows of matrix A and elements of vector c here.
#------------------------------------------------------------------
A=rbind(c(-1,4),c(3,6))    # Rows of A are -1,4; 3,6.
c=c(8,30)

#------------------------------------------------------------------
#  (a)  Solve system of linear equations.
#------------------------------------------------------------------
x=solve(A,c)

#------------------------------------------------------------------
#  (b)  Invert matrix A.
#------------------------------------------------------------------
Ainv=solve(A)

#------------------------------------------------------------------
#  Print results to the console.
#------------------------------------------------------------------
x
Ainv
Ainv%*%A      # Check inverse
```

Run the script, and the following will be printed at the console:

```
> x
[1]  4 3
> Ainv
             [,1]          [,2]
[1,]  -0.3333333 0.22222222
[2,]   0.1666667 0.05555556
> Ainv%*%A
       [,1]            [,2]
[1,]      1 1.665335e-16
[2,]      0 1.000000e+00
```

Note in the script that using two arguments in the `solve()` function in the form `solve(A,c)` produces the solution to the system of equations defined by $\mathbf{Ax} = \mathbf{c}$, whereas using one argument in the form `solve(A)` produces the inverse of the matrix $\mathbf{A}$. In the output, the inverse is calculated as decimal numbers with the unavoidable round-off error. You can see that calculating $\mathbf{A}^{-1}\mathbf{A}$ at the end produced *very nearly*, but not quite, an identity matrix, the discrepancy from a perfect identity being due to round-off error.

Like real numbers, not all square matrices have inverses. Unlike real numbers, there are many matrices without inverses. For instance, if we alter the 2×2 coefficient matrix in our example system of equations to be

$$\mathbf{A} = \begin{bmatrix} -1 & 4 \\ -2 & 8 \end{bmatrix},$$

then an inverse for $\mathbf{A}$ does not exist. The corresponding linear equations,

$$-x_1 + 4x_2 = 8,$$

$$-2x_1 + 8x_2 = 30,$$

turn out to be two lines with the same slope but with different vertical axis intercepts, that is, they are parallel and thus never meet. The second row of $\mathbf{A}$ is seen to be two times the first row, and so the two lines have some redundant properties in common (in this case, slope). The general math result is that if one of the rows of a $k \times k$ matrix is equal to the sum of the other rows (with each row is scalar-multiplied first by a constant, with the constants possibly different for each row), then neither an inverse for the matrix nor a point solution to a system defined by that matrix exist. A matrix with no inverse is said to be **singular**, and solving a system defined by such a matrix is sort of like dividing by zero. A matrix with an inverse is **nonsingular**.

Try feeding the equations with the revised singular matrix into R for a solution, and see what happens. Do it in a large open field, far away from people or buildings.

Real-World Examples

Old Faithful

At the Old Faithful Visitor Center, a few hundred meters from Old Faithful geyser in Yellowstone National Park, the park rangers post a prediction of the time when the next eruption of the geyser will take place, so that the crowds of visitors can gather at the right time. Although the average time until the next eruption is somewhere around 75 minutes, the actual times from eruption to eruption are quite variable. Old Faithful is fairly faithful for a geyser but not so faithful for impatient visitors on a tight schedule. How do the rangers know when the next eruption will be? Do they have some sort of valve for turning on the eruption, like turning on a slide show in the Visitors Center theater (Figure 12.2)?

FIGURE 12.2
The crowds gather at the right time for the next eruption of Old Faithful to begin.

The data in Table 12.1 consist of pairs of numbers recorded for a sample of eruptions of the geyser. Each row represents one eruption. The column labeled x contains the duration in minutes of the eruptions, and the column labeled y contains the amounts of time that elapsed until the next eruption. Take the time now to draw a scatterplot of the data, using R. Write a small script (hey, I cannot do *all* the work here) and obtain something like Figure 12.3.

It is evident from Figure 12.3 that there seems to be a strong positive relationship between the duration of an eruption and the time until the next eruption. In fact, the relationship is nearly linear. Long eruptions seem to deplete the underground reservoirs more, and so the time until the reservoirs are replenished and flash to steam from the underground volcanic heating is greater.

The rangers exploit the relationship between eruption duration and time until the next eruption for predicting the time until the next eruption. One might guess that there is a ranger out there in the crowds with a stopwatch!

In fact, a simple line of the form $y_{predicted} = b_1 + b_2 x$, where b_1 is an intercept and b_2 is a slope, that passes somehow through the middle of the data might offer a serviceable prediction. The ranger would just need to time the eruption, put the time into the equation as the value of x, and out would pop the predicted amount of time $y_{predicted}$ until the next eruption. But which line? Every different pair of values of b_1 and b_2 gives a different line—which line should the rangers use (Figure 12.4)?

TABLE 12.1

Observations of the Duration (x) in Minutes of an Eruption, and
Time (y) in Minutes until the Next Eruption, for a Sample of 35
Eruptions of Old Faithful Geyser in Yellowstone National Park

x	y
1.80	56
1.82	58
1.88	60
1.90	62
1.92	60
1.93	56
1.98	57
2.03	60
2.05	57
2.13	60
2.30	57
2.35	57
2.37	61
2.82	73
3.13	76
3.27	77
3.65	77
3.70	82
3.78	79
3.83	85
3.87	81
3.88	80
4.10	89
4.27	90
4.30	84
4.30	89
4.43	84
4.43	89
4.47	86
4.47	80
4.53	89
4.55	86
4.60	88
4.60	92
4.63	91

One of the most important and widely used applications of solving a system of linear equations is for finding a good prediction line in just such a situation in which there is a strong linear relationship between two quantities. We want to predict one quantity, called the response variable, from another quantity, called the predictor variable. This statement of the objective sparks

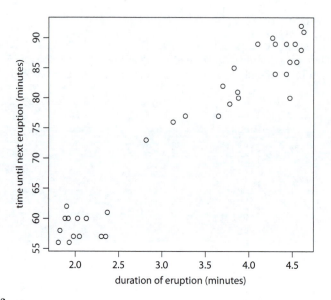

FIGURE 12.3
Observations on duration of eruption in minutes, and waiting time until the next eruption, from a sample of eruptions of Old Faithful geyser, Yellowstone National Park.

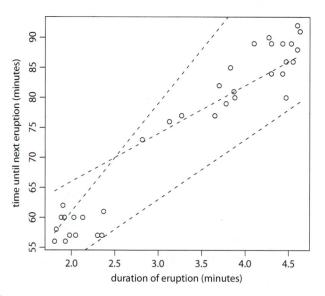

FIGURE 12.4
Dashed lines: three different pairs of intercept (b_1) and slope (b_2) values produce three different prediction lines given by the linear equation $y_{predicted} = b_1 + b_2 x$, where x is time until the next eruption in minutes and $y_{predicted}$ is the predicted amount of time until the next eruption. Circles: observations from a sample of eruptions of Old Faithful geyser, Yellowstone National Park.

the idea for finding a good predictor line: perhaps use the line which predicts the data in hand the best! To be more precise about what is meant by "best," we will have to specify some quantitative measure of how well a particular line predicts the data.

One quality in prediction that seems desirable is that small prediction errors are not bad, but big prediction errors are very bad. The Old Faithful crowds such as in Figure 12.2 might not mind waiting 10 or 20 minutes but might not care to wait an hour or more for the show to begin. This notion, that a lot of small prediction errors can be tolerated so long as large prediction errors are vastly reduced, can be conveniently summarized quantitatively using the concept of **squared prediction error** for each observation. For an observation, say the ith one, take the value y_i of the response variable, subtract the value $b_1 + b_2 x_i$ predicted for that observation by a particular line calculated with the value of the predictor variable, and square the result: $(y_i - b_1 - b_2 x_i)^2$. This squared prediction error is a measure of lack of fit for that observation. It agrees with our notion that such a measure should magnify a large departure of the observed value of y from the value predicted for y by the line.

We can use the sum of the squared prediction errors for all n observations in the data as the overall measure of how poorly a given line predicts the data. That quantity is the **sum of squared errors** (SSE) given by

$$SSE = \left(y_1 - b_1 - b_2 x_1\right)^2 + \left(y_2 - b_1 - b_2 x_2\right)^2 + \cdots + \left(y_n - b_1 - b_2 x_n\right)^2.$$

The criterion we can use is to pick the values of b_1 and b_2 that make SSE as small as possible. The values of b_1 and b_2 that minimize the sum of squared errors are called the **least squares estimates** of b_1 and b_2. The least squares criterion for picking prediction equations is used extensively in science and business.

The amazing thing is that there is usually a unique pair of values (let's call them $\hat{b}_1$, $\hat{b}_2$) that minimize the sum of squared errors for any given data set. In other words, one unique line minimizes SSE. Even more amazing is that we can find the least squares estimates of b_1 and b_2 with some straightforward matrix calculations involving a system of linear equations!

We will use the Old Faithful data to illustrate how the matrix calculations are set up. One would set up the calculations the same way with data on some other response and predictor variables. First, build a vector, let's call it **y**, that contains the observations on the response variable (the variable to be predicted). In the subsequent calculations, **y** will be treated as a column vector with n rows:

$$\mathbf{y} = \begin{bmatrix} 56 \\ 58 \\ \vdots \\ 91 \end{bmatrix}.$$

Next, build a matrix, let's call it **X**, with n rows and two columns, in which the elements of the first column are all 1s and the elements of the second column are the observations on the predictor variable:

$$\mathbf{X} = \begin{bmatrix} 1 & 1.80 \\ 1 & 1.82 \\ \vdots & \vdots \\ 1 & 4.63 \end{bmatrix}.$$

Then, form the transpose of **X**, usually denoted **X'**, which is a matrix in which the columns of **X** are formed into rows:

$$\mathbf{X'} = \begin{bmatrix} 1 & 1 & \cdots & 1 \\ 1.80 & 1.82 & \cdots & 4.63 \end{bmatrix}.$$

We will denote by **b** the vector of the unknown intercept and slope constants:

$$\mathbf{b} = \begin{bmatrix} b_1 \\ b_2 \end{bmatrix}.$$

The following remarkable result was first published (in nonmatrix form) over 200 years ago by the French mathematician Legendre in 1805 (see Stigler 1986).

Result. The least squares estimates of the intercept and slope constants are a point solution to the system of linear equations given by

$$\mathbf{Ab} = \mathbf{c},$$

where **A** is a square matrix formed by $\mathbf{A} = \mathbf{X'X}$ and **c** is a column vector formed by $\mathbf{c} = \mathbf{X'y}$.

From Legendre's result, all we have to do is create the matrices and solve the system! The solution, denoted $\hat{\mathbf{b}}$ if it exists, is given by $\mathbf{A}^{-1}\mathbf{c}$, or

$$\hat{\mathbf{b}} = \left(\mathbf{X'X}\right)^{-1}\mathbf{X'y}.$$

So, with a matrix inverse and a few matrix multiplications, we can predict Old Faithful. We can take advantage of the transpose function in R: t() takes a matrix as an argument and returns its transpose. Let us get to work:

```
#==================================================================
# R script to calculate intercept b[1] and slope b[2] for the
# least squares prediction line.  Script produces a scatterplot
# of the data along with overlaid prediction line.
```

```
#
# Data file should have two columns:  response variable to be
# predicted labeled "y" and predictor variable labeled "x".
# Change the R working directory to the location of the data file.
# Re-label axes in the plot() function.
#==============================================================

#-----------------------------------------------------
# Input the data.
#-----------------------------------------------------
Geyser=read.table("old_faithful.txt",header=TRUE) # Change file name
                                                  #  if necessary.
attach(Geyser)

#-----------------------------------------------------
# Calculate least squares intercept and slope.
#-----------------------------------------------------
n=length(y)              # Number of observations is n.
X=matrix(1,n,2)          # Form the X matrix:  col 1 has 1's,
X[,2]=x                  #  col 2 has predictor variable.
b=solve(t(X)%*%X,t(X)%*%y)   # Least squares estimates in b;
                             #  t( ) is transpose function.
                             # Alternatively can use
                             #   b=solve(t(X)%*%X)%*%t(X)%*%y.

#-----------------------------------------------------
# Draw a scatterplot of data, with superimposed least
# squares line.
#-----------------------------------------------------
plot(x,y,type="p",xlab="duration of eruption (minutes)",
 ylab="time until next eruption (minutes)") #Scatterplot of data.
ypredict1=b[1]+b[2]*min(x)  # Calculate predicted y values at
ypredict2=b[1]+b[2]*max(x)  # smallest and largest values of x.
ypredict=rbind(ypredict1,ypredict2)
xvals=rbind(min(x),max(x))
points(xvals,ypredict,type="l")  # Connect two predicted values
                                 #  with line.

#-----------------------------------------------------
# Print the intercept and slope to the console.
#-----------------------------------------------------
"least squares intercept and slope:"
b
```

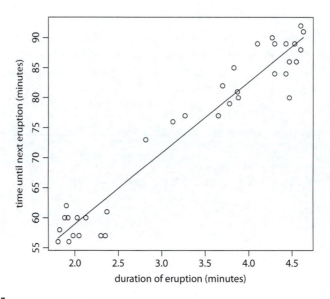

FIGURE 12.5
Circles: observations from a sample of eruptions of Old Faithful geyser, Yellowstone National Park. Line: least squares prediction line given by $y_{\text{predicted}} = 35.30117 + 11.82441x$.

The output at the console looks like this:

```
[1] "least squares intercept and slope:"
> b
          [,1]
[1,]  35.30117
[2,]  11.82441
```

Our prediction equation for Old Faithful is $y_{\text{predicted}} = 35.30117 + 11.82441x$. The script produces Figure 12.5. When issuing an actual prediction, the rangers usually give a window of time, centered at the predicted time from their model, to allow for the scatter of actual eruption times around the prediction.

A Galaxy Not So Far Away

If you know how bright a candle is at a known distance, and then you view the candle at a farther, unknown distance, you can calculate how far away the candle is by measuring its apparent brightness. Here are the details. The brightness decreases with distance according to a simple inverse square relationship. If L is the brightness (luminosity) of the candle at distance D, then the inverse square relationship is

$$L = \frac{c}{D^2},$$

where c is a constant that depends on the candle. (We will not derive the relationship here, but it is easy to envision. Think of the candle inside a huge ball and measure the amount of light falling on an area, say one square meter, on the inside of the ball. Now think of the situation with a ball twice as large. All the light from the candle is spread over a much larger surface area, and so the amount of light falling on a square meter is much less).

Now, suppose you measure L at known distance D. Then suppose the candle is removed to a farther unknown distance, let us call it d, and you measure the brightness to be l (naturally somewhat dimmer than L). Because $l = c/d^2$, the ratio of the brightness that you measured is

$$\frac{l}{L} = \frac{D^2}{d^2}.$$

The cs get cancelled; we only need to know the brightness l relative to L and the distance D to find the distance d. So, multiply both sides by d^2, divide both sides by l/L, and then take the square root of both sides to get the distance formula:

$$d = D\left(\frac{L}{l}\right)^{1/2}.$$

Everything on the right-hand side is measured, and so the unknown distance d can be calculated.

Likewise, for measuring distances to astronomical objects like stars and galaxies, if we only knew the luminosity (now we will switch to the astronomical term for brightness) of such an object at a known distance, we could easily calculate the distance to that object from its apparent luminosity.

Henrietta Leavitt documented a "standard candle" in 1912 that revolutionized astronomy and our conception of the size of the universe (see Johnson 2005). The Magellanic Clouds are small galaxies that are near, and satellites of, our own Milky Way galaxy. They are readily seen in the Southern Hemisphere (but not in the Northern) and are an extraordinary sight through binoculars on a dark clear night. Leavitt painstakingly measured the magnitudes of thousands of a particular type of recognizable star in the Small Magellanic Cloud using photographic plates taken through telescopes. The type of star is recognizable because it fluctuates in magnitude quite regularly over the course of some days. After some graphing, Leavitt discovered a strong near-linear relationship between the apparent magnitudes of the variable stars and the logarithms of their periods of fluctuation.

Here are data on nine variable stars in the Small Magellanic Cloud:

period	apparent magnitude
1.2413	17.21
1.6433	16.96
2.7120	16.54
2.8742	16.47
3.0655	16.31
3.1148	16.31
3.2139	16.56
3.9014	15.89
6.6931	15.69

Using a natural (base e) logarithm of a period in the role of x, and apparent magnitude in the role of y, the least squares prediction line for these nine stars is found to be $y_{predicted} = 17.4307669 - 0.9454438x$. You should repeat this calculation, adapting the script for the least squares line from the Old Faithful example. The data and prediction line appear in Figure 12.6.

The value of such a linear relationship was immediately recognized. It can be used to estimate the actual distance to astronomical objects, like star clusters and even galaxies, that contain this type of variable star!

These variable stars are common in our galaxy and other galaxies. They are called Cepheid variable stars after one of the first such stars discovered (Delta Cephei in the constellation Cepheus). Some Cepheids are close enough that one can determine their actual distance by parallax (the triangulation method studied in Chapter 9). With such a nearby star of known distance, one can feed its period into the linear relationship to predict how bright such a star would appear to us in the Small Magellanic Cloud—or in any other object, such as a star cluster or galaxy, that contained enough recognizable Cepheids for such a linear relationship to be determined reliably.

The distance to a Cepheid star in the Small Magellanic Cloud can then be estimated with the luminosity–distance equation. The stars in the cloud are much, much closer to each other than they are to us, so the distance to a star in the cloud is essentially the distance to the cloud itself.

The original Cepheid, Delta Cephei, is a magnitude 4.0 star visible to the naked eye. It was recently measured by the Hubble Space Telescope to be 273 parsecs (1 parsec is about 3.26 light years) away. Its period is 5.36634 days. So, if we could transport the star to the Small Magellanic Cloud, our linear relationship predicts that the star would have the following magnitude:

$$y_{predicted} = 17.4307669 - 0.9454438 \times \log(5.36634).$$

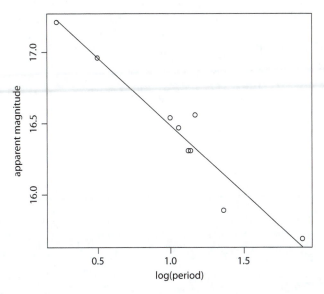

FIGURE 12.6

Circles: observations from a sample of classical Cepheid variable stars in the Small Magellanic Cloud. Line: least squares prediction line given by $y_{predicted} = 35.30117 + 11.82441x$.

Calculate it at the R console:

```
> x=log(5.36634)
> x
[1] 1.680146
> y.predicted=17.4307669-0.9454438*x
> y.predicted
[1] 15.84228
```

The value of $\log(5.36634)$ was printed so that we could see where it falls on our line (Figure 12.6). Of course, we cannot transport Delta Cephei to the Small Magellanic Cloud, but the line allows us to predict how bright an exactly identical Cepheid would be in the Small Magellanic Cloud. A predicted magnitude of 15.84228 is very dim indeed, but well within the range of telescopes even 100 years ago.

Now an important point is that the line predicts *magnitude*. To calculate distance, we need luminosity. Does something sound familiar? We have seen the relationship between luminosity and magnitude in Chapter 10. Magnitude is a logarithmic scale. Recall that magnitude is luminosity on a reverse logarithmic scale, using that weird base of $100^{1/5}$ for the logarithm. If M is magnitude, the relationship is

$$L = 100^{-M/5}.$$

Suppose M is the magnitude of the star at the known distance D, and suppose m is the value of its predicted magnitude in the Small Magellanic Cloud (which we calculated as y_{predict}). We substitute $L = 100^{-M/5}$ and $l = 100^{-m/5}$ into the distance formula:

$$d = D\left(\frac{L}{l}\right)^{1/2} = D\left(\frac{100^{-M/5}}{100^{-m/5}}\right)^{1/2} = D100^{(m-M)/10}.$$

We have everything on the right-hand side: $D = 273$ parsecs, $M = 4.0$, $m = 15.84228$. To the console:

```
> D=273
> M=4.0
> m=15.84228
> d=D*100^((m-M)/10)
> d
[1] 63770.33
```

The Small Magellanic Cloud is estimated to be 63770.33 parsecs or $63770.33 \times 3.26 = 207891.3$ light years away. Warp factor 9, Mr. Sulu, if we are to arrive before dinner.

Final Remarks

Our graphs have been quite functional, but they are a little plain. In Chapter 13, we will jazz them up with titles, legends, lines, text, and multiple panels. We will also take the opportunity to do some three-dimensional plotting.

Least squares can be used to incorporate multiple predictor variables for predicting a response variable. Least squares can also be used to fit curves to data for prediction. Although the material involved might be getting close to the level of university graduate science courses, we will try some out in Chapter 15.

WHAT WE LEARNED

1. A system of k linear equations in k unknowns given by

$$a_{11}x_1 + a_{12}x_2 + \cdots + a_{1k}x_k = c_1,$$
$$a_{21}x_1 + a_{22}x_2 + \cdots + a_{2k}x_k = c_2,$$
$$\vdots$$
$$a_{k1}x_1 + a_{k2}x_2 + \cdots + a_{kk}x_k = c_k,$$

can be represented in matrix form:

$$\mathbf{Ax} = \mathbf{c}.$$

Here, the element in the ith row and jth column of the matrix $\mathbf{A}$ is a_{ij}, $\mathbf{x}$ is a column vector containing the elements $x_1, x_2, ..., x_k$, and $\mathbf{c}$ is a column vector containing $c_1, c_2, ..., c_k$.

2. The identity matrix, a matrix with 1s on the main diagonal (upper left to lower right) and 0s elsewhere, usually denoted $\mathbf{I}$, is the multiplicative identity of matrix multiplication: $\mathbf{CI} = \mathbf{C}$ and $\mathbf{IC} = \mathbf{C}$, where $\mathbf{C}$ is any matrix.

3. The inverse of a square matrix $\mathbf{A}$, if it exists, is a square matrix usually denoted $\mathbf{A}^{-1}$, with the following property:

$$\mathbf{AA}^{-1} = \mathbf{A}^{-1}\mathbf{A} = \mathbf{I}.$$

4. A system of k linear equations in k unknowns given by $\mathbf{Ax} = \mathbf{c}$ has a unique point solution given by

$$\mathbf{x} = \mathbf{A}^{-1}\mathbf{c},$$

provided the inverse $\mathbf{A}^{-1}$ exists.

5. In R, the function `solve(A, c)`, where A is a $k \times k$ matrix and c is a vector with k elements, returns the solution $\mathbf{A}^{-1}\mathbf{c}$ if a solution to the system given by $\mathbf{Ax} = \mathbf{c}$ exists. The function `solve(A)` returns the inverse matrix $\mathbf{A}^{-1}$, if it exists.

Example:

$$\mathbf{A} = \begin{bmatrix} 3 & 4 & -6 \\ 2 & -5 & 17 \\ -1 & 8 & -4 \end{bmatrix}, \quad \mathbf{c} = \begin{bmatrix} 10 \\ 30 \\ 5 \end{bmatrix}$$

```
> A=rbind(c(3,4,-6),c(2,-5,17),c(-1,8,-4))
> c=c(10,30,5)
> solve(A,c)
[1]  4.288889 2.100000 1.877778
> solve(A)
              [,1]         [,2]          [,3]
[1,]   0.25777778 0.07111111 -0.08444444
[2,]   0.02000000 0.04000000  0.14000000
[3,]  -0.02444444 0.06222222  0.05111111
```

6. With observations consisting of recorded values $y_1, y_2, ..., y_n$ of a variable to be predicted (response variable) and corresponding values $x_1, x_2, ..., x_n$ of a variable (predictor variable) to be used to predict the response variable, the estimated intercept $\hat{b}_2$ and the estimated slope $\hat{b}_2$ of the least squares prediction line are the respective elements in the vector **b** in the system of linear equations given by

$$(\mathbf{X'X})\mathbf{b} = \mathbf{X'y},$$

where prime ' denotes matrix transpose, and

$$\mathbf{X} = \begin{bmatrix} 1 & x_1 \\ 1 & x_2 \\ \vdots & \vdots \\ 1 & x_n \end{bmatrix}, \ \mathbf{y} = \begin{bmatrix} y_1 \\ y_2 \\ \vdots \\ y_n \end{bmatrix}$$

The solution, if it exists, is given by

$$\mathbf{b} = (\mathbf{X'X})^{-1} \mathbf{X'y}.$$

Computational Challenges

12.1. Calculate the inverses of the following matrices:

$$\begin{bmatrix} -0.8 & 3.9 \\ -3.4 & 9.1 \end{bmatrix} \quad \begin{bmatrix} -0.02 & -4.7 & 4.9 \\ -6.6 & -2.7 & 0.7 \\ 3.9 & -0.8 & -3.7 \end{bmatrix}$$

$$\begin{bmatrix} 1 & 1 & 1 & 1 \\ 10 & 7.9 & -12.6 & 2.7 \end{bmatrix} \quad \begin{bmatrix} 1 & 10 \\ 1 & 7.9 \\ 1 & -12.6 \\ 1 & 2.7 \end{bmatrix}.$$

12.2. Solve the systems of linear equations of the form $\mathbf{Ax} = \mathbf{c}$ for the vector of unknowns, corresponding to the following **A** and **c** matrices:

$$
\begin{bmatrix}
14.5 & 5.6 & -5.1 \\
10.2 & -8.7 & -3.5 \\
7.5 & 10.0 & 2.2
\end{bmatrix},
\begin{bmatrix}
17.9 \\
10.4 \\
-3.8
\end{bmatrix}.
$$

$$
\begin{bmatrix}
3.6 & -7.5 & 3.0 & -6.9 \\
-13.3 & 8.2 & -6.7 & 17.1 \\
.7 & -12.7 & -2.4 & 17.8 \\
-3.2 & 25.9 & -17.9 & -11.4
\end{bmatrix},
\begin{bmatrix}
6.1 \\
1.1 \\
4.6 \\
6.7
\end{bmatrix}.
$$

12.3. Here are some of the winning times (seconds) in the men's 1500 m race in the Olympics, 1900–2008. We saw these data earlier in Chapter 7. Develop a linear prediction equation for the trend of the winning times through the years. Does the trend in recent years look like it is continuing according to the linear model, or changing? Issue a prediction for the winning time in 2012, obtain the real winning time with some web research, and compare the prediction to the real thing.

```
1900   1920   1936   1948   1960   1968   1972   1976   1980
246.0 241.9 227.8 229.8 218.4 214.9 216.3 219.2 218.4

1984   1988   1992   1996   2000   2004   2008
212.5  16.0 220.1 215.8 212.1 214.2 212.9
```

12.4. Below are data for the 50 states plus the District of Columbia on per capita income (PCI) and percent college graduates (PCG). Develop a prediction equation to predict PCI from PCG and plot data and equation together on a graph.

```
PCI   37036 31121 34539 29330 24820 31276 35883 26874 36778
PCG    31.7  27.6  26.6  22.9  22.4  21.1  24.5  18.8  22.5

28158 31107 28513 33219 35612 29136 27215 27644 25318 37946
 23.8  24.3  21.0  26.0  25.5  22.3  15.3  25.1  20.1  35.5

30553 30267 35409 32103 38390 38408 33327 28061 31899 29387
 23.4  28.0  29.9  25.9  33.1  35.4  34.2  30.8  28.1  25.5

37065 32478 33616 32462 31252 33116 32836 31614 28352 41760
 26.9  24.6  24.8  24.5  24.2  24.4  30.0  25.5  24.9  35.2

33565 43771 40507 36153 47819 37373 31395 32315 36120 34897
 25.6  34.6  30.6  27.2  34.5  32.5  25.2  24.4  27.4  25.3

44289 54985
 36.7  45.7
```

12.5. Newmark (1996) summarized extinctions of mammal species in parks and nature reserves in Tanzania. For each of six parks, he calculated the extinction rate λ from the exponential decay equation:

$$S = S_0 e^{-\lambda t},$$

where S is the current number of mammal species in the park, S_0 is the number of mammal species in the park when it was first established as a protection area, and t is the amount of time that has elapsed since the park was first established. Each park had its own value of λ, and a park's value of λ was strongly related to the logarithm of the area of the park. Your job is to reproduce Newmark's calculations. On the next page are the raw data consisting of four variables: A is area of the park, t is time (years) since the establishment of the park, S is present number of species, and S_0 is initial number of species when the park was established.

Calculate the extinction rate λ for each park (figure out how to do this from the exponential decay equation). Calculate $\log(A)$ for each park. Draw a scatterplot to see if $\log(A)$ might be a good predictor of extinction rate λ. If the plot looks promising, calculate the slope and intercept for the least squares prediction line. Draw a graph with a scatterplot of the data and the least squares prediction line superimposed.

A	t	S_0	S
100	36	35	33
137	35	26	25
1834	83	23	21
2600	38	41	40
12950	44	39	39
23051	55	49	49

Afternotes

1. Matrices, solutions of systems of linear equations, and matrix inverses are some of the important subjects contained in the branch of mathematics known as linear algebra.

2. The first electronic computer, called the Atanasoff–Berry Computer (ABC; see http://www.youtube.com/watch?v=YyxGIbtMS9E), was completed in 1942 and was designed solely for the task of solving systems of linear equations. It could handle up to 29 equations, a task for which it would need about 25 hours. You can handle, say, 2900 equations without too much trouble using one statement in the R console:

```
> x=solve(matrix(runif(2900*2900),2900,2900),runif(2900))
```

The statement generates a 2900×2900 matrix of random numbers between 0 and 1 in the role of the coefficient matrix **A**, and 2900 random numbers between 0 and 1 in the role of the vector **c**. Be patient; go get a cup of coffee. The calculations could take a minute or so. When the console prompt > appears, the calculations are done and stored as the vector x, with 2900 elements. Printing the results would take more time. Can your puny earthling computer handle 29,000 equations?

3. A closer inspection of the Old Faithful data (Figure 12.3) reveals what seem to be two clusters of data points. The geyser appears to have two "personalities": a regime in which the eruptions are long and the intervals between eruptions are long, and a regime in which the eruptions are short and the intervals between eruptions are short. The situation resembles what dynamic systems modelers would call a system with "two alternative locally stable states." Think of two shallow bowls next to each other on a rickety table, with a marble in one of the bowls. The table shakes randomly with every passing person, and the marble wobbles around in the bowl. An occasional large shake propels the marble over the rim and into the other bowl, where it wobbles around until propelled back again. Somehow the underground plumbing of Old Faithful could have two different steady states, either burping up small amounts of water incessantly or gushing large amounts of water less frequently. Some random perturbation pushes Old Faithful from one state to the other. The math of the situation is beyond the scope of this book. If you are interested, however, some college calculus and some coursework in dynamic systems modeling would equip you to further explore the topic. I should say that not only is the math beyond the scope of the book but also you might have to contribute to the mathematical model-building efforts once you are equipped: the workings of Old Faithful and other geysers are at present not well understood.

References

Johnson, G. 2005. *Miss Leavitt's Stars: The Untold Story of the Woman Who Discovered How to Measure the Universe.* New York: W. W. Norton.

Newmark, W. D. 1996. Insularization of Tanzanian parks and the local extinction of large mammals. *Conservation Biology* 10:1549–1556.

Stigler, S. M. 1986. *History of Statistics: Measurement of Uncertainty Before 1900.* Cambridge: Harvard University Press.

13

Advanced Graphs

Here, we will look at some more graphical resources in R. First, we will study some of the more commonly used ways to customize R graphs. We will concentrate mainly on plots of data and models: points, curves, and so on. The customizations, if appropriate, are readily applicable to other types of graphs such as boxplots and histograms. Second, we will try out some techniques for multiple-panel graphs and for plotting three-dimensional data.

Two-Dimensional Plots

The two main functions in R for two-dimensional plotting (horizontal and vertical axes) are plot() and matplot().

The plot() function produces standard x–y graphs in Cartesian coordinates. Its takes two vectors as arguments: the first being the values for the horizontal axis and the second being the values for the vertical axis. Example (script or console):

```
x=(0:100)*2*pi/100
y=sin(x)
plot(x,y)
```

The plot() function will work with just one vector argument of y-values, in which case an index of x-values will be automatically generated by R. For example,

```
x=(0:100)*2*pi/100
y=sin(x)
plot(y)
```

The matplot() function takes two matrices as arguments. The columns of one matrix are plotted against the corresponding columns of the second matrix, on the same graph. The number of rows for the matrices should match. If the rows do not match, the shorter matrix should be augmented with missing values (NA code). The first matrix argument contains the values for the horizontal axis. If one of the matrices has only one column, that column is used in all the plots. Different colors are used for each plotted column pair, and numbers are used as the plotting symbols for points. These aspects can be overridden with various options such as `col=` and `pch=`; see options below. The `matplot()` function saves the trouble of specifying multiple `points()` statements for adding points to a plot, and the default axes are automatically sized to include all the data.

For example,

```
x=(0:100)*2*pi/100
y1=sin(x)
y2=cos(x)
y=cbind(y1,y2)
matplot(x,y)
```

Options for Styles of Symbols, Lines, Axes

Many options exist for customizing the display of R graphs. Usually, options are invoked as additional arguments in the `plot()` or `matplot()` functions. The optional arguments can be entered in any order (after the vectors to be plotted), separated by commas. Most of the options have default settings that will be in force if you omit those arguments. The default settings used by R are usually pretty good, but there is frequently room for improvement. We have used some of these options already and will now review them more systematically.

Data Symbol Types

Symbols used for the data points in a plot are designated with the `pch=` option (pch for "plotting character"); for instance, `pch=1` would draw data points as circles. The common symbols and their numerical codes appear in Figure 13.1.

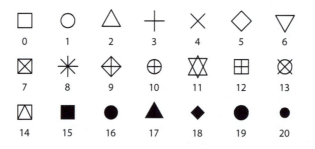

FIGURE 13.1

Symbol types and their numerical codes for drawing data points, specified with the `pch=` option.

Connecting Line Types

If a line is used to connect data points or portray curves, the style of the line can be controlled with the `lty=` option. The common line types and their numerical codes appear in Figure 13.2.

Incidentally, when using the `lty=` and other options in `matplot()`, setting the option to a vector value will cause the multiple plots to rotate sequentially through the list of different styles specified. For example,

```
x=(0:100)*2*pi/100
y1=sin(x)
y2=cos(x)
y=cbind(y1,y2)
matplot(x,y,type="l",lty=c(1,2),col="black")
```

The `col="black"` option as used above suppresses the use of different line colors in the `matplot()` function.

Plot Types

Different plot types are specified with the `type=" "` option; for instance, `type="p"` specifies a plot with only points drawn and no connecting lines. The common types are as follows:

p: Only symbols for points drawn; no connecting lines.

l: Only connecting lines drawn; no symbols for points.

o: Both symbols for points and connecting lines drawn.

b: Same as o, except that the connecting lines do not touch the point symbols.

h: Vertical line from each point to horizontal axis ("comb" plot).

s: "Staircase" plot, with neighboring points connected by a horizontal line followed by a vertical line.

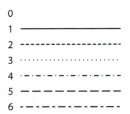

FIGURE 13.2
Line types and their numerical codes for drawing data points, specified with the `lty=` option.

Axis Limits

The limits for the horizontal and vertical axes are specified respectively with the `xlim=` and the `ylim=` options. For instance, `xlim=c(0,10)` forces R to use a horizontal axis with a range from 0 to 10. If data are outside that range, they will not appear on the plot.

Tic Marks

The tic marks on the axes are controlled with the `lab=` and `tcl=` options. The setting of the `lab=` option is a vector with two elements: the first being the number of tic marks on the horizontal axis and the second being the number of tic marks on the vertical axis. For instance, `lab=c(7,3)` gives an *x*-axis divided into eight intervals by seven tic marks and a *y*-axis divided into four intervals by three tick marks. R might override the `lab=` option if the choices do not work well in the display. The `tcl=` option gives the length of the tic marks as a fraction of the height of a line of text. The setting of `tcl` can be a negative number, in which case the tic marks will be on the outer side of the axes; a positive setting (default) produces tic marks on the inner side of the axes.

Axis Labels

The labels for the horizontal and vertical axes are the names of the plotted vectors by default. The labels can be changed with the `xlab=" "` and the `ylab=" "` options, where the desired text strings are included between the quotes. For instance, `xlab="time"`, `ylab="population size"`.

Suppressing Axes

Sometimes figures are better without axes. For instance, Figure 13.1 is just all the plotting characters drawn on a grid of points in an *x–y* coordinate system (along with text characters added to the graph for the numerical codes). The axes are simply omitted from the figure because they contribute no useful information. Axes are suppressed with the option `axes=FALSE`, and the axis labels (and titles) are suppressed with the option `ann=FALSE`.

Sizes of Symbols, Labels, Widths of Lines, Axes

The size of plotting characters can be altered with the `cex=` option. The numerical value specified for `cex` is a multiplier of the default character size. The default value of `cex` is 1. For instance, `cex=2` plots the data symbols at twice the normal size; `cex=.5` yields half the normal size. Sizes of other plot aspects are controlled similarly as multiplier values through the following options:

`cex.axis=` Axis width multiplier

`cex.lab=` Axis label text size multiplier

`cex.main=` Main title text size multiplier (title specified in title() function)

`cex.sub=` Subtitle text size multiplier

`lwd=` Plotting line width multiplier

Other Customizations

Other customizations for R graphs are not arguments in the plotting functions but rather require additional statements.

Adding Points

Additional data or model curves can be added to an existing, open graph with the `points()` or `matpoints()` functions. The `points()` function adds to graphs created with `plot()`, and the `matpoints()` function adds to graphs created with `matplot()`. The two functions otherwise are used like the `plot()` and the `matplot()` functions, with many of the arguments for the appearance of data and curves working exactly the same. Options that attempt to alter aspects of the existing graph, such as the limits of axes, will not work, however.

Adding Lines

Lines can be added to an existing, open graph with the `lines()` and `matlines()` functions. The functions take two vectors as arguments: the first gives the *x*-coordinates and the second vector gives the *y*-coordinates of the points to join with lines. A line is drawn between each successive pair of points given by the vectors. The optional arguments governing line types and styles can be used.

Disjunct (nonconnecting) line segments can be easily added to an existing, open graph with the `segments()` function. An alternative is to invoke multiple `lines()` commands. The `segments()` function takes four vector arguments. The first two vectors give, respectively, the *x*- and *y*-coordinates of the points at which the line segments are to originate, and the second two vectors give the *x*- and *y*-coordinates of the points where the segments are to terminate. Successive different plotting styles for the segments can be specified (in this function and others) by assigning vector values for the plotting style options. An example use of the `segments()` function would be

```
x0=c(2,2,3,4)
y0=c(2,3,2,2)
x1=c(2,3,3,4)
y1=c(4,3,4,4)
x=c(0,3)
```

```
y=c(0,3)
plot(x,y,type="l",lty=0,xlim=c(0,5),ylim=c(0,5))  # Draws axes
                                                  # with
                                                  # blank plot
                                                  # region.
segments(x0,y0,x1,y1)
```

Lines drawn across the entire graph can be added with the abline()
function. The function can take the following three forms:

abline(a,b) a is the intercept, b is the slope, of a single line to be
  drawn.

abline(h=) Draws a horizontal line at the *y*-value specified by h=.

abline(v=) Draws a vertical line at the *x*-value specified by v=.

An example of the use of abline() would be

```
x=(0:100)*2*pi/100
y=sin(x)
plot(x,y,type="l",lty=1)
abline(h=0,lty=2)
```

Adding Text

Text can be added to the plotting region with the text() function. The
text() function takes two vector arguments, giving the sets of *x–y* coordinates where the text items are to be located. An additional character vector
argument gives the respective text strings to be drawn. The pos= option has
possible values 1, 2, 3, 4, giving the position of the text in relation to the coordinate point. Omitting pos= has the effect of overlaying the text centered on
the point. The following example draws Figure 13.3:

```
x=c(2.5,1,1,4,4)
y=c(2.5,1,4,4,1)
plot(x,y,type="p",xlim=c(0,5),ylim=c(0,5))
text(x[2:5],y[2:5],c("bottom","left","top","right"),pos=1:4)
text(x[1],y[1],c("overlay"))
```

Titles and Subtitles

Plot titles can be added with the title() function:

```
x=c(1,2,3)
y=c(3,3,3)
plot(x,y)
title(main="wow what a plot")
```

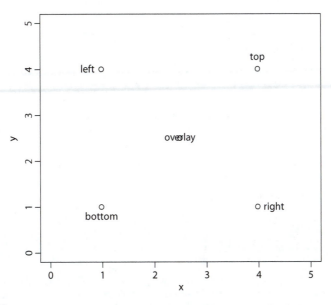

FIGURE 13.3
Graph that shows different text positions in relation to coordinate points specified by the pos= option in the text() function. pos=1: bottom, pos=2: left, pos=3: top, pos=4: right. When the pos= option is omitted, the text is overlaid on the coordinate point.

The same effect can be achieved with the main= option within the plot() function (and other graphing functions):

```
x=c(1,2,3)
y=c(3,3,3)
plot(x,y,main="wow what a plot")
```

If the text argument in either case is a character vector, then a title with multiple lines will be produced:

```
x=c(1,2,3)
y=c(3,3,3)
plot(x,y)
title(main=c("wow","what a plot"))
```

A subtitle, or text below the plot, can be added with the sub= option as follows:

```
x=c(1,2,3)
y=c(3,3,3)
plot(x,y)
title(main="wow what a plot",sub="and the labels rock too")
```

The sub= option works as well in the plot() function. Invoking the title() function gives more flexibility, such as when an overall title is desired for a graph with multiple panels.

Legends

Legends can be added to plots with the legend() function. The first two arguments are the *x*- and *y*-coordinates of the upper left of the legend box. The next arguments give the text strings for the legend entries and sets of plotting symbols that are to appear in the box, such as pch= and/or lty=. It often helps to draw the plot first and then look for a nice open area to locate the legend. For example,

```
x=(0:100)*2*pi/100
y1=sin(x)
y2=cos(x)
y=cbind(y1,y2)
matplot(x,y,type="l",lty=c(1,2),col="black")
legend(3,1,c("sine","cosine"),lty=c(1,2))
```

New Graph Windows

A new graph window is opened in R with the following commands:

windows() (in Windows)

quartz() (in Mac OS)

X11() (in Unix, Linux)

You can use the appropriate command as a line or lines in a script when producing more than one graph. Otherwise, each successive graph will replace the previous one.

Global and Local

When options are invoked in the plot(), matplot(), or other plotting functions, they are local to that function call. All the default settings are resumed for other plots drawn in the same work session. The par() function (plotting parameters function) can be used to fix display settings globally for a sequence of plots. For instance, one might want to change the text size or fix the line types for all the plots. The par() function accepts most of the optional arguments that are normally used in the plotting functions. When an option is subsequently changed in a plotting function, the change remains in effect only for that plot. The option then reverts to whatever was specified in par().

Multiple Panels

A figure can be built with multiple panels, each panel having a different plot. The construction is accomplished with the `layout()` function. The main argument to the layout function is a matrix containing the integers 1, 2, ..., k, where k is the number of panels to be drawn. The matrix indicates the position in the figure of each panel. The first plot drawn is in panel 1, the second plot drawn is in panel 2, and so on. Remember that by default, the `matrix()` function reads in data by columns. Thus, the panel that I chose to label "c" in the figure is drawn second. The following script produces the four quadratic functions in Figure 8.5:

```
#  The four quadratic functions in Figure 8.5.
xhi=2
xlo=-1
x=xlo+(xhi-xlo)*(0:1000)/1000      # Range of x values from xlo
                                   # to xhi

# Set the layout of the panels
layout(matrix(c(1,2,3,4),2,2))     # First row is 1, 3.  Second
                                   # row is 2, 4.

#  Plot 1.
y=x^2-x-1                                # Calculate quadratic.
plot(x,y,type="l",ylim=c(-2,2))          # Plot quadratic.
text(-.75,1.75,"a")                      # Label the panel.
y2=numeric(length(x))                    # Vector of zeros.
points(x,y2,type="l",lty="dashed")       # Plot the zero line.
                                         # lty="dashed" is
                                         # same as lty=2.


# Plot 2.
y=-x^2+x+1
plot(x,y,type="l",ylim=c(-2,2))
text(-.75,1.75,"c")
points(x,y2,type="l",lty="dashed")


# Plot 3.
y=x^2-x+1
plot(x,y,type="l",ylim=c(-2,2))
text(-.75,1.75,"b")
points(x,y2,type="l",lty="dashed")


# Plot 4.
y=-x^2+x-1
plot(x,y,type="l",ylim=c(-2,2))
text(-.75,1.75,"d")
points(x,y2,type="l",lty="dashed")
```

Scatterplot Matrices

A remarkable form of a multiple-panel plot is called a **scatterplot matrix**, sometimes abbreviated as **SPLOM** in statistics books. A scatterplot matrix is simply an array of scatterplots of every pair of quantitative variables in a data set. The display is handy for exploring potential relationships in complex data. In R, a scatterplot matrix can be produced easily with the ever-versatile plot() function. All that one needs to do is use a data frame as the argument of the plot statement, instead of vectors! All or most of the variables in the data frame should be quantitative; R will turn categorical variables into quantitative ones for the plotting, but the results for such variables might be nonsensical.

Let us try an example of a scatterplot matrix. We will need some data. R has a way of continually surprising a user with its abundance of resources, and it turns out that R is filled with ready-to-use data frames containing interesting data. At the console, type data() to see a list of the data frames that are automatically loaded when you turn on R. The data are used in statistics courses to illustrate various analysis techniques. You will note in the list, for instance, an enlarged data set of Old Faithful timings. We will use the famous iris data, consisting of measurements of the lengths and the widths of flower parts from samples of flowers among three different species of irises. You can see the data just by typing iris at the console.

Just type the following command at the console:

```
> plot(iris)
```

The result is Figure 13.4. In each individual scatterplot, you can perceive different clusters of points. The clusters correspond to the different species; this type of plot is a jumping-off point in taxonomy for numerically discriminating between similar species. An improved plot would identify and/or color points associated with individual species. However, that identification is perhaps best done on larger scatterplots. R turned the species' categorical variable into numbers for plotting; those plots in the scatterplot matrix are meaningless.

Three-Dimensional Plots

Producing good three-dimensional plots can be a challenge, but such plots can be quite informative. Here, we will look at two ways of depicting three-dimensional information in R: surface (wire mesh) plots and contour plots.

The surface plot or wire mesh plot works with a grid of *x*–*y* values. Over each value of *x* and *y*, a value, for example, *z*, is recorded or calculated. The *z*-values can be thought of as heights in a landscape recorded at coordinates

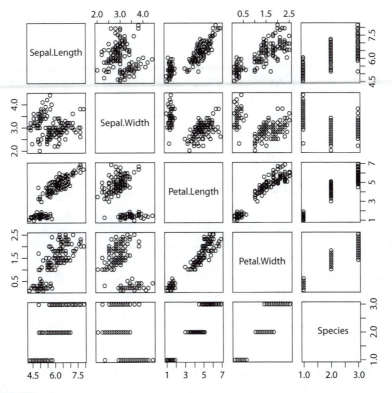

FIGURE 13.4
Scatterplot matrix of data consisting of sepal length, sepal width, petal length, and petal width measurements for flowers from three iris species.

given by the combinations of x and y values. The plot links the adjacent z-values with lines, as if the landscape is being covered with a flexible screen or wire mesh.

An example of a numerical landscape is the sum of squares surface we have encountered before when fitting a prediction line to the data. For a given data set, the sum of squared errors was minimized to find the intercept and slope corresponding to the "best" prediction line. Each different pair of intercept and slope values gives a different sum of squares value. We can visualize the least squares estimates of intercept and slope as the location of the lowest point in a "valley," with the elevation at any location in the valley given by the sum of squares value.

Here is a script to illustrate the use of the persp() function (perspective function), which draws a landscape data set as a wire mesh plot. The persp() function takes three arguments: a vector of x-coordinates, a vector of y-coordinates, and a matrix of z-values with elements calculated at every x-y pair. We will look at the sum of squares surface for the Old Faithful data

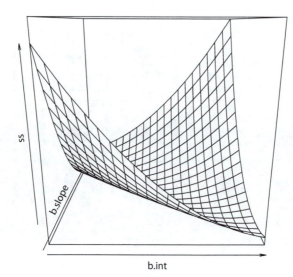

FIGURE 13.5
Sum of squares surface produced by different intercept and slope values of prediction lines for
the Old Faithful data of Chapter 12.

from Chapter 12. In our case, the *x*-coordinates are a range of intercept values
approximately centered at the least squares estimate of 35.3. The script puts
these intercept values in the vector b.int. The *y*-coordinates are a range
of slope values centered approximately at the least squares estimate of 11.8.
These slope values are put in the vector b.slope.

A for-loop picks out each element of b.int in turn for calculating the sum
of squares. Inside that for-loop, another "nested" for-loop picks out each element of b.slope. In this way, each pair of intercept and slope values is singled out for the calculation of the sum of squares. The sum of squares values
are stored in the matrix ss. The vectors b.int and b.slope and the matrix
ss become the arguments in the persp() function.

For any three-dimensional graph, one normally must try different ranges
and mesh sizes to capture the important landscape details. Run this script,
which produces Figure 13.5. Remember to set your working directory to
whatever folder contains the old _ faithful.txt data from Chapter 12.
Afterwards, feel free to experiment with different ranges for b.int and
b.slope:

```
#===========================================================
# R script to draw 3d surface (wire mesh plot).
# Example is sum of squares surface for slope and intercept
# of prediction line for Old Faithful data.
#===========================================================
```

```
#---------------------------------------------------------
# Input the data.
#---------------------------------------------------------
Geyser=read.table("old_faithful.txt",header=TRUE) # Change
                                                   # file name
                                                   # if
                                                   # necessary.
attach(Geyser)
#---------------------------------------------------------
# Set up X matrix.
#---------------------------------------------------------
n=length(y)               # Number of observations is n.
X=matrix(1,n,2)           # Form the X matrix:  col 1 has 1's,
X[,2]=x                   #   col 2 has predictor variable.
#---------------------------------------------------------
# Calculate range of intercept (b.int) and slope (b.slope)
# values.  Use 21X21 grid.
#---------------------------------------------------------
b.int=(0:20)*10/20+30     # Range from 30 to 40.
b.slope=(0:20)*4/20+10    # Range from 10 to 14.
#---------------------------------------------------------
# Calculate matrix (ss) of sum of squares values.  Rows
# correspond to intercept values, columns correspond to
# slope values.
#---------------------------------------------------------
ss=matrix(0,length(b.int),length(b.slope))
for (i in 1:length(b.int)) {
  for (j in 1:length(b.slope)) {
    b=rbind(b.int[i],b.slope[j]) # Col vector of intercept,
                                 # slope.
    ss[i,j]=sum((y-X%*%b)^2)     # Sum of squares;  X%*%b is
                                 # col vector
                                 # of predicted values.

  }
}
#---------------------------------------------------------
# Draw the sum of squares surface.
#---------------------------------------------------------
persp(b.int,b.slope,ss)   # R function for drawing wire mesh
                          # surface.
detach(Geyser)
```

The sum of squares landscape in Figure 13.5 looks to be a long and narrow valley with a nearly flat valley bottom. The least squares estimates of intercept and slope produce a sum of squares value only slightly lower than neighboring locales along the valley bottom. We can depict the situation in an alternative way, with a contour plot. Think of what a topographic map of the sum of squares valley might look like: that map is a contour plot.

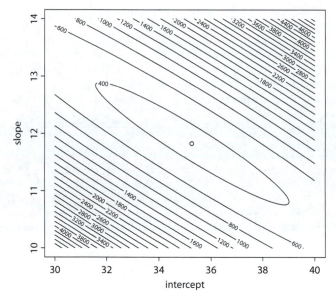

FIGURE 13.6

Contour plot of the sum of squares surface produced by different intercept and slope values of prediction lines for the Old Faithful data of Chapter 12.

The function `contour()` in R draws contour plots. It takes the same type of data arguments as `persp()`. The following script is mostly the same as the previous one for the wire mesh plot. However, calculating the landscape with a finer mesh helps define the contours better. The last part of the script calculates the least squares estimates and locates them on the landscape. The script produces Figure 13.6:

```
#===========================================================
# R script to draw contour plot of 3-d surface.
# Example is sum of squares surface for slope and intercept
# of prediction line for Old Faithful data.
#===========================================================
#-----------------------------------------------------------
# Input the data.
#-----------------------------------------------------------
Geyser=read.table("old_faithful.txt",header=TRUE)  # Change
                                                   # file name
                                                   # if
                                                   # necessary.

attach(Geyser)
#-----------------------------------------------------------
# Set up X matrix.
#-----------------------------------------------------------
```

```
n=length(y)                # Number of observations is n.
X=matrix(1,n,2)            # Form the X matrix:   col 1 has 1's,
X[,2]=x                    #   col 2 has predictor variable.


#------------------------------------------------------------
# Calculate range of intercept (b.int) and slope (b.slope)
# values.   Use 201X201 grid.
#------------------------------------------------------------
b.int=(0:200)*10/200+30      # Range from 30 to 40.
b.slope=(0:200)*4/200+10     # Range is from 10 to 14.
#------------------------------------------------------------
# Calculate matrix (ss) of sum of squares values.   Rows
# correspond to intercept values, columns correspond to
# slope values.
#------------------------------------------------------------
ss=matrix(0,length(b.int),length(b.slope))
for (i in 1:length(b.int)) {
  for (j in 1:length(b.slope)) {
    b=rbind(b.int[i],b.slope[j]) # Col vector of intercept,
                                 # slope.
    ss[i,j]=sum((y-X%*%b)^2)     # Sum of squares;   X%*%b is
                                 # col vector
                                 # of predicted values.

  }
}
#------------------------------------------------------------
# Draw the sum of squares contour plot.   Calculate least
# squares solution for intercept and slope and add solution
# point to plot.
#------------------------------------------------------------
contour(b.int,b.slope,ss,nlevels=30,xlab="intercept",
  ylab="slope")
bhat=solve(t(X)%*%X,t(X)%*%y)
points(bhat[1],bhat[2])
detach(Geyser)
```

Color

Color should be used judiciously and sparingly in data graphics. When color is used, it should convey important information and not just be there for, uh, color. The problem is that color is a distraction. When symbols in a graph have different colors, the eye is searching for what the meaning of the color is. Color purely as decoration takes the attention away from the informational point that the graph is supposed to convey. A secondary but important problem with color in data graphics is that a small but appreciable fraction of humans are color blind.

That being said, colors in the plotting functions are controlled by a collection of options:

col= Colors of data symbols and lines in plot() and matplot(); colors of bars in barplot()

col.axis= Colors of axes

col.lab= Colors of axis labels

col.main= Colors of main titles

col.sub= Colors of subtitles

Colors can be specified with numerical codes; for instance, col=25 produces blue. Type colors() at the console to see 657 colors recognized by R. More conveniently, all the color names listed in the 657 colors are text strings that are recognized by R in color options. So, col="blue" produces the identical effect to col=25.

Final Remarks

The essence of R is graphics. We have learned here some powerful tools for visualization of quantitative information. Yet we have barely scratched the surface of the graphical resources available in R. However, the hardest part about learning R is getting started in the basics. Once you have a sense of the concepts and structure of R, you will find it easy as well as fascinating to explore the vast wealth of graphical displays in R that are available, either inside R itself or as R scripts and packages contributed by scientists around the world. Murrell's (2006) excellent book is highly recommended for further study of the graphical capabilities of R.

WHAT WE LEARNED

1. The main functions in R for two-dimensional Cartesian graphs are plot(), which plots vectors, and matplot(), which plots columns of matrices.
2. Many alterations to graphs are entered as optional arguments in plotting functions like plot() and matplot(). Some of the often used options are as follows:
   pch= Plotting characters
   lty= Line types
   type=" " Type of plot
   xlim=, ylim= Limits to *x*- and *y*-axis
   xlab=" ", ylab=" " Labels for *x*- and *y*-axis

`main=" "` Plot title.

`sub=" "` Plot subtitle.

`lab=` Number of tic marks on x- and y-axis.

`tcl=` Length and orientation of tic marks.

`axes=FALSE` Suppresses drawing of axes.

`ann=FALSE` Suppresses drawing of axis labels and titles.

`cex=` (and `cex.axis=`, `cex.lab=`,`cex.main=`,`cex.sub=`) Size multiplier of plotting characters, axis widths, axis labels, title text, and subtitle text.

`lwd=` Line width multiplier.

`col=` (and `col.axis=`, `col.lab=`, `col.main=`, `col.sub=`) Colors of plotting symbols and lines, axes, axis labels, titles, and subtitles.

3. Additional modifications to graphs are accomplished with additional functions:

   `points()`, `matpoints()` Adds points to plots made with `plot()` and `matplot()` functions.

   `lines()`, `matlines()` Connects points with lines on plots made with `plot()` and `matplot()` functions.

   `segments()` Adds line segments to plots.

   `abline()` Adds a horizontal, vertical, or line with designated intercept and slope to a plot.

   `text()` Adds text to a plot.

   `title()` Adds a title to a plot.

   `legend()` Adds a legend to a plot.

   `windows()`, `quartz()`, `X11()` Opens a new graphics window (Windows, Mac OS, Unix/Linux).

   `par()` Set graphical options globally for ensuing plots.

4. Figures with multiple panels can be produced with the following functions:

   `plot()` Produces scatterplot matrix (SPLOM) when argument is a data frame.

   `layout()` Divides figure into an array of plot regions, for multiple panels.

5. Three-dimensional information in the form of numerical landscapes can be portrayed with surface (or wire mesh) plots and with contour plots. Some functions in R for drawing such plots are as follows:

   `persp()` Surface (wire mesh) plot. Arguments are vector of x-coordinates, vector of y-coordinates, and matrix of z-values.

   `contour()` Contour plot. Arguments are vector of x-coordinates, vector of y-coordinates, and matrix of z-values.

Computational Challenges

13.1. Listed below are 8 years of data collected by the U.S. Bureau of Land Management on the pronghorn (*Antilocapra americana*) population in the Thunder Basin National Grassland in Wyoming. The variables are fawn count y, pronghorn population size u, annual precipitation v, and winter severity index w. Draw a scatterplot matrix of these variables:

Y	u	v	w
290	920	13.2	2
240	870	11.5	3
200	720	10.8	4
230	850	12.3	2
320	960	12.6	3
190	680	10.6	5
340	970	14.1	1
210	790	11.2	3

13.2. Add a legend to Figure 7.2.

13.3. The data set volcano is one of the data sets available in R. Type volcano at the console to view the numbers. The data set is simply a matrix of elevations of the region around the Maunga Whau volcano in New Zealand. Draw a surface plot and a contour plot of the volcano.

13.4. Draw a surface plot of the following function:

$$z = \frac{1}{2\pi\sqrt{1-\rho^2}} \exp\left[x^2 + 2\rho xy + y^2\right]$$

Here, x and y are coordinates, and ρ is a constant with value between -1 and $+1$. Use values of x between -2 and $+2$, values of y between -2 and $+2$, and use the value $\rho = 0.8$. Once your script is up and running, repeat the plot for the following values of ρ: 0.9, 0.6, 0.2, -0.2, -0.6, -0.9.

13.5. Draw contours of the function from Computational Challenge 13.4, under the different values of ρ given.

13.6. For the data in Computational Challenge 2.8, plot the data for all three species of lady beetles on one graph. Use different symbols for the data from different species of lady beetles. Add the fitted kill rate curves for each species. Provide a legend.

13.7. Add two vertical dashed lines to Figure 9.12 at horizontal axis values -1 and $+1$ in order to illustrate the eccentricity of Earth's orbit. Label the orbit curve as Earth's orbit. Label perihelion and aphelion (the periapsis and the apsis of solar orbits). Draw a small circle at the origin and label it as the sun. Provide a title.

13.8. Create a multiple-panel figure like Figure 4.1 to illustrate different ways of graphing data for a single numerical variable (stripchart, histogram, boxplot, and timeplot). Choose or find a vector of data from the list of data sets in R, or from elsewhere, that would be suited to this illustration.

Reference

Murrell, P. 2006. *R Graphics*. Boca Raton, FL: Chapman & Hall/CRC Press.

14

Probability and Simulation

We encountered some probability and simulation earlier in this book, in Chapter 7. There, we simulated many sequences of 30 at-bats of a baseball player. We saw that according to the model, the outcomes could vary substantially simply by chance: some weeks, the player will be on a tear, other weeks the player will be in a deep slump. While such hot and cold streaks are consistent with athletes' experiences, it is surprising that a simple unchanging success–failure random mechanism could produce such wide swings in performance.

Random Variables

In the baseball simulation, we recorded the proportion of successes (hits) out of 30 attempts (at-bats). When we repeated the process (simulated another 30 at-bats), we typically saw a new proportion occur. The proportion of successes was a quantity likely to vary if the process being observed was repeated.

Quantities that vary at random are called, appropriately enough, random variables. Many processes generate random variables. Trap a bird at random from a population of a finch species and measure its beak length. Trap another bird from the population and you will likely have a bird with a slightly different beak length. Here, the random variable is the beak length. Draw a student at random and record the number of hours the student has spent at a social media web site in the past 24 hours. Draw another student and record that student's 24-hour social media time. The numerical result for each student is likely different. The random variable is the number of hours of social media time.

In each case above, the process generating the data is complex and multi-layered. The biological processes produced variation in beak lengths, and the sampling process picks the actual bird that contributes a beak length to the data. Likewise, for the social media data, the variability of social media times among students is produced by myriad forces at work in students' lives, and the sampling mechanism singles out the actual data value recorded.

In science, random variables are important because they are typically counts or measurements produced by some processes by which scientific data are generated. Science is filled with quantities that vary: amount of daily rainfall at a weather station, household income among households in a city, growth yields of field plots of wheat under a particular fertilizer treatment, distances of galaxies from Earth, the number of craters in 100-square-kilometer areas on the surface of Venus, and survival times of patients with a particular form of cancer under a particular form of treatment.

What can scientists possibly learn when a process that generates counts or measurements gives a new number practically every time the process is repeated? The answer is that science has devised ways of looking for *patterns in the variability*.

The study of ways to make reliable conclusions from variable counts and measurements is the branch of science called **statistics**. The heart of statistics lies in building and testing mathematical models of how the counts and measurements arise, for use as hypotheses about our understanding of the processes involved. The study of patterns in randomness is a branch of mathematics called **probability**.

Probability

A deep theorem from mathematics states the following: If a random process could be repeated many, many times, the long-run proportion (or fraction) of times that a particular outcome happens stabilizes. The theorem is known as the **law of large numbers**. For a small number of repetitions, as in repeating a baseball at-bat only 30 times, we can expect considerable variability in the proportion of hits. However, for a large number of repetitions, say 1000 at-bats (two seasons of an active major league player), the long-run proportion of hits becomes less and less variable. While a player's short-run batting average can jump around, the long-run batting average settles into a near-fixed constant practically tattooed on the player's forehead. The player can change that number only by substantially altering the batting mechanics and the approach to hitting, that is, by becoming somehow a different data generation process.

Let us simulate this idea. In our simulation from Chapter 7, the player undergoes a sequence of trials called "at-bats." During each at-bat, we classify the outcome as either a success (hit) or a failure (such as an out or error). (The definition of "at-bat" in baseball does not include walks or being hit by the pitch, etc.) We recall that during any given at-bat, we had assumed that the probability of a success was .26. What does this probability really mean?

If the player has 10 at-bats, the possible outcomes are 0, 1, 2, 3, …, 10 hits (or successes). The fractions of successes corresponding to those outcomes would be 0, .1, .2, .3, …, 1. Interestingly, none of these fractions is equal to the probability .26. A success proportion of .26 is not even a possible outcome of these 10 at-bats! Furthermore, from our simulation in Chapter 7, we saw that over a sequence of 30 at-bats, the player's observed batting average (proportion of successes) might turn out to be over .400, or it might turn out to be under .100, or anything in between! A success probability of .26 is deeply hidden if not invisible in the random process that we observe.

Yet the success probability is there, lurking behind the outcomes. We can begin to find it with an R script. Using R, we can perform all kinds of experiments that might not be available to us in life. For instance, we can ask our baseball player to have 100 at-bats, or 500 at-bats, or even every number of at-bats between 5 and 500!

Recall that our outcomes() function that we wrote in Chapter 7 produces a vector of 0s and 1s representing random failures and successes. This time, instead of repeating blocks of 30 at-bats, we will try a block of 5 at-bats, then a block of 6 at-bats, then a block of 7, and so on. We have some serious computational horsepower in R at our disposal, and so we can do every number of at-bats up until, say, 500. For each block, we will calculate a batting average. Then, we will plot the batting average and the number of at-bats to see how fast the variability in the batting average decreases.

Try this script:

```
#==========================================================
# Simulation of the law of large numbers.
#==========================================================

#----------------------------------------------------------
# 1.  Function to generate sequence of successes and
# failures.  Arguments:  number of trials is n, success
# probability on any given trial is p.  Function returns
# vector x of 0's and 1's.
#----------------------------------------------------------
outcomes=function(n,p) {
  u=runif(n)
  x=1*(u<=p)    # R converts logical vector to numeric 0's
                #   and 1's in attempting to multiply by 1.
  return(x)
}

#----------------------------------------------------------
# 2.  Simulate every number of trials from 5 to max.n.
# Store proportions of successes in p.obs.
#----------------------------------------------------------
p=.26            # Probability of success.
max.n=500        # Maximum number of trials.
n=5:max.n
p.obs=numeric(length(n))
for (i in 1:length(n)) {
  p.obs[i]=sum(outcomes(n[i],p))/n[i]
}

#----------------------------------------------------------
# 3.  Plot p.obs versus n.
#----------------------------------------------------------
plot(n,p.obs,type="l",lty=1,xlab="number of trials",
  ylab="proportion of successes")
abline(h=p,lty=2)
```

Part 1 of the script just rebuilds our function called `outcomes()` from Chapter 7 for generating a random vector of 0s and 1s. Here, the function is altered a bit from its Chapter 7 version: although it still outputs the random vector of 0s and 1s, I changed the for-loop inside the function into a vector calculation using a cute trick. Look at the statement `x=1*(u<=p)`. Recall that u is a vector of random numbers between 0 and 1, and p is the success probability on any given trial. The expression `u<=p` returns a logical vector with elements `TRUE` or `FALSE`. Then, `1*(u<=p)` attempts to multiply the logical vector by 1, an operation that would normally make no sense, as a logical vector contains no numbers to multiply. However, when R encounters this expression, R automatically changes the logical vector into a string of 0s and 1s so that the arithmetic can be performed (the R jargon is that R "coerces" the logical vector into a numeric vector). The expression `(u<=p)+0` would accomplish the same coercion and the same result.

"Vectorizing" calculations instead of using loops makes R run faster. However, remember for your own work that excessively compact scripts can be hard to debug. Cute is sweet. Too cute can be chaos.

Part 2 of the script produces two vectors: n contains the integers from 5 to 500, and p.obs contains the observed proportion of successes for our `outcome()` function calculated for every value in n. The two vectors are plotted in Part 3. The script produces Figure 14.1.

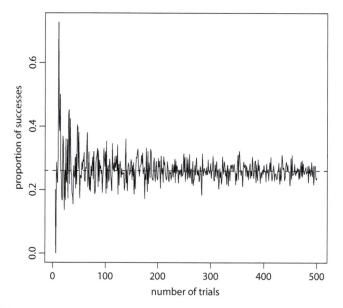

FIGURE 14.1
Observed proportion of successes in a random success–failure process plotted with number of trials. Dashed line indicates probability of success on any given trial.

Note in Figure 14.1 how the variability of the observed proportion of successes narrows as the number of trials (at-bats) increases. The observed proportion is rarely if ever equal to the underlying success probability, but the observed proportion becomes better and better predicted by the underlying success probability.

The notion can be turned around to provide a working definition of the probability of an event for scientific use. The probability of an event is defined as the quantity that the observed proportion of times the event occurs converges to as the number of trials increases, each trial ending in either the event occurring or the event not occurring. In practice, an event probability might never actually be known but rather must be estimated.

Probability Distributions of Counts

There are regularities in randomness. The regularities come from the long-run stable proportions of outcomes.

This time, instead of keeping track of batting average, we will count the number of successes. We will simulate 30 at-bats over and over again, say, 10,000 times. As before, we assume that any practice effect is negligible so that the player's chance of success during any given at-bat does not change. The outcome on each set of 30 at-bats will be a count with possible values 0, 1, 2, ..., 30. Of course, we might not expect 30 hits to occur very often, but we recognize that the event is possible.

We will draw a histogram of the frequencies of the hit count outcomes. A small alteration of the previous script will serve. We can use our outcomes() function intact. We further need to count successes (sum(outcomes(n,p)), store the counts in a vector (hits), and draw a histogram of the vector.

Try this script:

```
#===========================================================
# Simulation of the probability distribution of success
# count.
#===========================================================

#-----------------------------------------------------------
# 1.  Function to generate sequence of successes and
# failures.  Arguments:  number of trials is n, success
# probability on any given trial is p.  Function returns
# vector x of 0's and 1's.
#-----------------------------------------------------------
outcomes=function(n,p) {
  u=runif(n)
  x=1*(u<=p)
  return(x)
}
```

```
#-------------------------------------------------------------
# 2.   Simulate n trials num.sets times.
# Store counts of successes in hits.
#-------------------------------------------------------------

n=30
p=.26
num.sets=10000
hits=numeric(num.sets)
for (i in 1:num.sets) {
   hits[i]=sum(outcomes(n,p))
}


#-------------------------------------------------------------
# 3.   Plot histogram of hits.
#-------------------------------------------------------------
bounds=(0:31)-.5
hist(hits,breaks=bounds,freq=FALSE)
```

In Part 3 of the script, the `freq=FALSE` option has the histogram rectangle areas calculated as proportions (relative frequencies) instead of absolute frequencies. It is useful to prespecify the boundaries of the histogram intervals to be half-integers –.5, .5, 1.5, 2.5, …, 30.5. Then, each integer outcome has its own interval bin in which to accumulate frequencies. The desired boundaries are calculated and put in the vector `bounds` and then specified to the histogram in the `breaks=bounds` option.

The script produced Figure 14.2. Slightly different results will occur each time the script is run, but the basic patterns will be the same. The most likely outcomes are from 5 to 10, and out of 10,000 simulations of 30 at-bats, values of 17 or more hits occurred too rarely to appear on the graph.

With 10,000 simulations, the proportions (areas of the rectangles) in Figure 14.2 are close to their long-run stable values that would result from the law of large numbers. The rectangle area over each integer represents something close to the probability that the particular integer will be the outcome of 30 at-bats.

The collection of the possible outcomes (0, 1, 2, …, 30) of the random variable (number of hits out of 30 at-bats) along with the probabilities of the outcomes (rectangle areas) is called the **probability distribution** of the random variable.

Binomial Distribution

In Figure 14.2, we simulated a type of probability distribution called a **binomial distribution**. A random process that produces a binomial distribution has the following characteristics: (1) A collection of n trials takes place, with each trial having only two possible outcomes, one traditionally labeled "success" and the other one labeled "failure." (2) For each trial, the

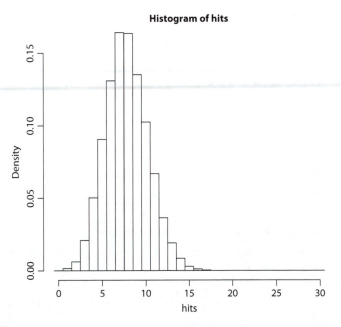

FIGURE 14.2
Relative frequency distribution from 10,000 simulations of a binomial distribution with $n = 30$ trials and success probability $p = .26$.

probability of success, denoted p, is the same. (3) The trials are **independent**, that is, the outcome of one trial does not affect the success probability for any other trial. (4) The random variable is the count of the number of successes and can take possible values 0, 1, 2, 3, …, n.

Calling one of the two outcomes of a trial a "success" and the other a "failure" is not a value judgment. The terms are just labels to help us keep track of which event is which throughout the complex analyses and the writing of lengthy scripts. In some studies, "success" might be something like "cancer" or "car crash," with "failure" being "no cancer" or "no car crash."

Let Y denote the random variable with a binomial distribution. Y is the count of successes in n trials, and its value will vary every time the whole process of n trials is repeated. Let k be a particular numerical outcome of the random variable; k in the formulas is a placeholder for one of the numbers 0, 1, 2, …, 30. The probability that Y takes the value k is denoted as $P(Y = k)$, and it is those probabilities for $k = 0, 1, 2, …, 30$ that are approximated by the histogram rectangles in Figure 14.2.

Remarkably, there is a formula for calculating these probabilities. Its derivation is not difficult, but it requires some preliminary study of basic probability mathematics, and so we will not derive it here. However, that will not stop us from using the formula and from using R to build simulations of binomial processes.

The binomial probabilities are given by the following formula:

$$P(Y = k) = \frac{n!}{k!(n-k)!} p^k (1-p)^{n-k},$$

for $k = 0, 1, 2, \ldots n$.

In the formula, $n!$ is read as "n factorial" and means $n(n-1)(n-2)\ldots(1)$. For instance, $4! = 24$.

The formula is remarkable in that it gives not just one probability distribution but rather many probability distributions. Each different pair of values of n and p produces a different set of probabilities.

R has several built-in functions to calculate various things related to the binomial distribution:

dbinom(k,n,p): Calculates $P(Y = k)$ from the above binomial probability formula. Here, k is the outcome, n is the number of trials, and p is the probability of success. Any of the arguments can be a vector.

pbinom(m,n,p): Calculates the sum of the binomial probabilities from $k = 0$ to $k = m$:

$$P(Y = 0) + P(Y = 1) + \ldots + P(Y = m) = P(Y \le m).$$

This corresponds to the probability that there would be m or fewer successes in n trials. Any of the arguments can be a vector.

rbinom(size,n,p): Simulates observations from a binomial distribution. Here, size is the length of the vector of binomial random variables, n is the number of trials for each binomial random variable, and p is the probability of success. This R function duplicates part of our script that produces Figure 14.2 (but it was helpful for our understanding to develop the calculations from scratch as we did).

Let us try these binomial functions. At the console, type:

```
> p=.26
> n=30
> k=0:30
> dbinom(k,n,p)
 [1] 1.193855e-04 1.258388e-03 6.410975e-03 2.102338e-02
 [5] 4.985950e-02 9.109465e-02 1.333593e-01 1.606490e-01
 [9] 1.622772e-01 1.393732e-01 1.028348e-01 6.569302e-02
[13] 3.654544e-02 1.777886e-02 7.585191e-03 2.842738e-03
[17] 9.363749e-04 2.709384e-04 6.875163e-05 1.525641e-05
[21] 2.948198e-06 4.932634e-07 7.089904e-08 8.664513e-09
```

```
[25]  9.363749e-04 2.709384e-04 6.875163e-05 1.525641e-05
[25]  9.913030e-15 2.402039e-16 2.813199e-18
> m=10
> pbinom(m,n,p)
[1]  0.8682599
> size=20
> rbinom(size,n,p)
 [1] 10  3 11  9  7 11  7  4  8  8  8  9  4  9  6  8  9  8 11  3
> probs=dbinom(k,n,p)
> probs=cbind(0,probs)
> k=cbind(-1,k)-.5
> plot(k,probs,type="s")
```

The dbinom(k,n,p) statement calculated all the binomial probabilities for our 30 at-bats model from 0 hits to 30 hits.

The pbinom(m,n,p) statement calculated the probability that the player would have 10 or fewer hits in 30 at-bats.

The last four statements produce the "staircase" plot in Figure 14.3. Compare with the histogram of the observed proportions out of 10,000 simulations in Figure 14.2. The staircase needs to rise at every half integer −.5, 1.5, 2.5, ..., 29.5 by an amount equal to the corresponding element in probs. It takes a little thought to line up the staircase properly. The statement probs=cbind(0,probs) attaches a 0 to the beginning of probs, and

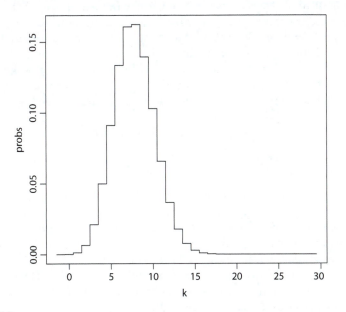

FIGURE 14.3
Staircase plot of binomial probabilities calculated for $n = 30$ trials and probability of success $p = .26$.

the statement k=cbind(-1,k)-.5 alters the values of k for plotting to be −1.5, −.5, 1.5, 2.5, ..., 29.5.

The variety of situations that could be modeled by binomial probabilities is vast—not just the number of hits out of any number of at-bats for any major league baseball player, but the number of elk that survive a winter out of *n* elk, the number of Obama supporters in a sample of *n* voters, the number of sample plots that contain ferns out of *n* sample plots randomly thrown in a forest, the number of test tubes displaying bacteria growth out of *n* test tubes of growth media inoculated with a diluted solution containing bacteria, and the number of atomic particles that decay within a period of time out of *n* particles.

Probability Distributions of Measurements

Life is discrete. Every measurement device by which we record data, whether it is a ruler or a spectrometer, only measures to a certain limit of accuracy, such as to the nearest second decimal place. The set of possible measurements that can be output by the device is a discrete set, meaning that we can count the members of the set. Even the set of numbers that can be stored in the floating point system of your computer is a discrete set (albeit a very large one).

Members of the set of real numbers in an interval, say, between 0 and 1, cannot be "counted" like the members of a discrete set. Real numbers are a mathematical abstraction beyond our everyday experience. Yet, there are numerous circumstances in science in which modeling the numerical output of some phenomenon as a set of real numbers offers great calculation convenience at little expense of realism. In probability, if the range of output of a random process is very "fine-grained" with many, many possible values, we can often achieve great simplifications by using a **continuous probability distribution** to model the process. A random variable with a continuous probability distribution (sometimes called a continuous random variable) takes a real value within a range of possible real values.

Uniform Distribution

When a random variable has a continuous distribution, we assign probabilities to intervals instead of individual numbers. For instance, we have already used a continuous distribution called the **uniform distribution**. In the uniform distribution we used, the random variable *U* conceptually takes a real value between 0 and 1. In our computer, however, *U* takes a value from any of the possible 16-decimal place numbers between 0 and 1. The conceptual uniform distribution has the following property: the probability that *U* takes a value between *a* and *b* is *b* − *a*, where $0 \le a \le b \le 1$:

$$P(a \le U \le b) = b - a.$$

For instance, the probability that U takes a value between .3 and .4 is $.4 - .3 = .1$. Certainly, calculating this probability with such a simple formula is easier than trying to simulate it with a script that loops 10,000 times (maybe if you are really getting into R you might disagree).

The probability that a continuous random variable takes a value between a and b is an *area* under what is called a **density curve**. The density curve is the theoretical histogram that would be approached with massive numbers of simulations of the random variable.

For the uniform random variable U, the density curve is just a flat line of height 1 over the interval 0 to 1. The quantity $b - a$ is then just the area under the flat line between a and b. More complex continuous distributions typically require computers (or extensive tables in the backs of statistics books) for calculating the areas. The entire area under a density curve is 1, just like the rectangle areas sum to 1 in a relative frequency histogram.

One seemingly odd consequence of using a continuous distribution can be seen from the uniform probability formula above. The probability that U takes any particular real number value, say r, is 0. Think of r being a number between a and b and then scrunch the interval by letting a approach r from below and b approach r from above. The width of the interval goes to 0. Remember that when working with real numbers, the event $U = r$ means that U equals r to the umpty-millionth decimal place. When working with real-life measurements, the event of recording a data value of .324 means that the actual value is somewhere between 3.235 and 3.245.

We would use the uniform as a model of discrete measurements when (A) the range of measurements is very fine-grained and (B) a flat line of height 1 over the interval 0 to 1 offers an adequate approximation to a histogram of the measurements.

Random numbers that have a uniform distribution on the interval 0 to 1 are simulated in R with the `runif()` function:

`runif(size)`: Simulates a vector of length `size` of uniform random variables on the interval between 0 and 1.

What the uniform distribution gives us is a convenient way to do many kinds of probability simulations and randomizations. We have already used the uniform distribution to simulate random vectors of 0s and 1s corresponding to a binomial distribution. Suppose there are 10 presentations to be given in your class and you want to have them go in a random order. Give the presentations unique identification numbers from 1 through 10. At the console, type:

```
> n=10
> id.order=rank(runif(n))
> id.order
 [1]   4 10  5  2  1  9  3  8  6  7
```

Here, n is the number of presentations. A vector of uniform random numbers of length n is produced by runif(n). Then, rank(runif(n)) returns the ranks of the random numbers (rank of 1 for the smallest, rank of 2 for second smallest, etc.). The result is a list of the integers from 1 to n in random order. And so, presentation 4 goes first; 7 is last. Type the statements again and you will get a different random order!

Normal Distribution

One of the most amazing results of probability is the **central limit theorem**. The central limit theorem states that sums of random variables from almost any probability distribution have probabilities that can be approximately calculated with areas under a particular bell-shaped curve. The bell-shaped curve corresponds to a continuous distribution known as the **normal distribution**.

Let us illustrate the concept of the central limit theorem first with an R script:

```
#================================================================
# Simulation of central limit theorem.
#================================================================

layout(matrix(c(1,2,3,4),2,2,byrow=TRUE))

#----------------------------------------------------------------
# One uniform random variable simulated 10000 times.
#----------------------------------------------------------------
size=1                      # Number of random variables in sum.
repeats=10000               # Number of values to simulate for
                            #  histogram.
v=runif(size*repeats)       # Vector of uniform random variables.
w=matrix(v,size,repeats)    # Enter v into a matrix (sizeXrepeats).
y=colSums(w)                # Sum the columns.
hist(y,freq=FALSE,ann=FALSE)         # Histogram.
title("size 1")

#----------------------------------------------------------------
# Sum of two uniform random variables simulated 10000 times.
#----------------------------------------------------------------
size=2                      # Number of random variables in sum.
repeats=10000               # Number of values to simulate for
                            #  histogram.
v=runif(size*repeats)       # Vector of uniform random variables.
w=matrix(v,size,repeats)    # Enter v into a matrix (sizeXrepeats).
y=colSums(w)                # Sum the columns.
hist(y,freq=FALSE,ann=FALSE)         # Histogram.
title("size 2")

#----------------------------------------------------------------
# Sum of five uniform random variables simulated 10000 times.
#----------------------------------------------------------------
```

```
size=5                          # Number of random variables in sum.
repeats=10000                   # Number of values to simulate for
                                #  histogram.
v=runif(size*repeats)           # Vector of uniform random variables.
w=matrix(v,size,repeats)        # Enter v into a matrix (sizeXrepeats).
y=colSums(w)                    # Sum the columns.
hist(y,freq=FALSE,ann=FALSE)            # Histogram.
title("size 5")

#----------------------------------------------------------------
# Sum of twenty uniform random variables simulated 10000 times.
#----------------------------------------------------------------
size=20                         # Number of random variables in sum.
repeats=10000                   # Number of values to simulate for
                                #  histogram.
v=runif(size*repeats)           # Vector of uniform random variables.
w=matrix(v,size,repeats)        # Enter v into a matrix (sizeXrepeats).
y=colSums(w)                    # Sum the columns.
hist(y,freq=FALSE,ann=FALSE)            # Histogram.
title("size 20")
```

The script produces four histograms, in which the random variable Y is the sum of 1, 2, 5, and 20 uniform random variables respectively. For each histogram, Y is simulated 10,000 times. The method is to fill a matrix called w that has size rows and 10,000 columns with uniform random numbers and then sum the columns of w with the colSums() function. The resulting vector y with 10,000 elements contains the values of the random variable Y for the histogram. The script produced Figure 14.4.

The successive histograms in Figure 14.4 approach an approximate symmetric bell shape as the number of random variables in the sum gets larger.

The central limit theorem provides the approximate bell shape in the form of an equation for a curve. The equation is

$$f(y) = \frac{1}{\sigma\sqrt{2\pi}} e^{-\frac{(y-\mu)^2}{2\sigma^2}},$$

and a graph of it appears in Figure 14.5. Here, μ is a constant that gives the center of the curve, and σ is a constant that measures the width of the curve (it is the horizontal distance from the center of the curve to the "inflection point," the place where the curve stops "taking a right turn" and starts "taking a left turn"). You might be able to tell from inspecting the formula that the height of the normal density curve on a logarithmic scale is a parabola (a quadratic function of y is in the exponent). The constants μ and σ differ from application to application, depending on the quantities being modeled. The central limit theorem further states that the approximation improves as the number of random variables in the sum increases.

The central limit theorem details will not be developed here, but they are a standard part of a high school or college-level introductory statistics

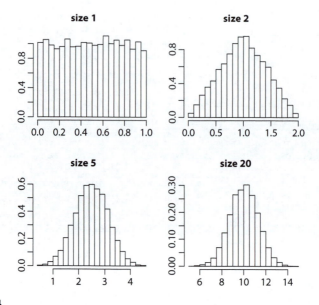

FIGURE 14.4
Histograms of 10,000 simulations of size 1, a single uniform random variable; size 2, a sum of 2 uniform random variables; size 5, a sum of 5 uniform random variables; and size 20, a sum of 20 uniform random variables.

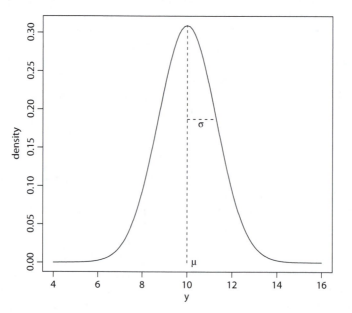

FIGURE 14.5
The density curve of the normal distribution approximation to the histogram (size 20) in Figure 14.4. The centering constant value is $\mu = n/2$, and the spread constant value is $\sigma = \sqrt{n/12}$, where n is the number of uniform random variables in the sum (here, $n = 20$).

course. (Proving the central limit theorem in full generality is remarkably difficult and is the subject of graduate-level mathematics courses!). Here, we can simply state that the central limit theorem produces the following values of μ and σ when n uniform random variables are being summed: $\mu = n/2$, $\sigma = \sqrt{n/12}$. The normal curve that approximates the histogram (size $n = 20$) from Figure 14.4 is drawn in Figure 14.5.

The point of the central limit theorem is that many different types of quantities measured in science are themselves sums of smaller, random quantities. SAT test scores are the sum of scaled positive or negative points for the correct or incorrect answers to questions. Baseball batting averages are the sums of contributions of small physical attributes and training histories of players. The yearly heights of a river on May 1 are the sums of many physical characteristics of the watershed and the region's local precipitation episodes during that winter and spring. Human heights (among males separately or among females separately) are the sums of small genetic and environmental contributions to bone lengths. However, because the μ value for male heights is different than that for female heights, combined heights of males and females do not usually display a bell-shaped histogram (but instead often a histogram with two peaks).

Even a process with a discrete binomial distribution has a good normal distribution approximation when the number of trials n is large. A binomial random variable is the sum of random 1s (successes) and 0s. The theoretical central limit theorem approximation for the binomial takes $\mu = np$ and $\sigma = \sqrt{np(1-p)}$.

Thus, we can frequently use the normal distribution to model the variability in many different types of random quantities in nature.

The normal distribution in R is implemented in the following functions:

dnorm(y,mu,sigma): Height of the normal density curve (bell-shaped curve) over y, having center constant mu and spread constant sigma.

pnorm(y,mu,sigma): The area under the normal density curve to the left of y.

rnorm(size,mu,sigma): Vector of length size of normally distributed random variables, from a distribution with center constant mu and spread constant sigma.

Let us try out these normal distribution functions. We can use SAT scores as an example to model. Nationally, SAT scores from a single part (such as the math part or the verbal part) are scaled so that μ is about 500 and σ is about 100. Go to the R console:

```
> mu=500
> sigma=100
> y=600*(0:100)/100+200      # Range of values from 200 to 800.
> density=dnorm(y,mu,sigma)   # Normal density curve.
> plot(y,density,type="l")
```

The resulting plot is the normal curve that approximates the histogram of the nation's SAT-math scores.

What is the approximate proportion of SAT math scores that are below 600? We can use the area under the normal curve to the left of 600 to find this proportion. Continue the console session:

```
> y=600
> pnorm(y,mu,sigma)
[1]  0.8413447
```

An approximate proportion .84, or 84%, of the scores are below 600. The score of 600 is called the 84th percentile of the SAT scores. (The values of μ and σ for SAT scores change a bit from year to year, and so the actual percentiles for that year might be a little different.) The total area under a normal density curve is 1, so that the proportion of scores above 600 can be calculated as

```
> 1-pnorm(y,mu,sigma)
[1]  0.1586553
```

The proportion of scores above 600 is around .16. What proportion of scores are between 400 and 700? This proportion can be calculated as the area to the left of 700 minus the area to the left of 400:

```
> pnorm(700,mu,sigma)-pnorm(400,mu,sigma)
[1]  0.8185946
```

A rule of thumb that students in statistics courses learn is that in a normal distribution, around 68% of the values in the histogram being approximated are between $\mu - \sigma$ and $\mu + \sigma$, and about 95% of the values are between $\mu - 2\sigma$ and $\mu + 2\sigma$:

```
> y1=mu-sigma
> y2=mu+sigma
> pnorm(y2,mu,sigma)-pnorm(y1,mu,sigma)
[1]  0.6826895
> y1=mu-2*sigma
> y2=mu+2*sigma
> pnorm(y2,mu,sigma)-pnorm(y1,mu,sigma)
[1]  0.9544997
```

If y is a vector containing data observations that has a reasonably bell-shaped histogram, then μ and σ for an approximating normal distribution can usually be adequately estimated respectively by mean(y) (the arithmetic or sample mean of the elements in y) and std(y) (the sample standard deviation of the elements of y). The sample standard deviation of the elements in y is alternatively written in "R-ese" as `sqrt(sum((y-mean(y))^2)/` `(length(y)-1))`. It is a measure of how spread out the elements of y are around their arithmetic mean.

Real-World Example

The normal distribution is frequently used in simulating the future of changing quantities. One key application is the calculation of risks associated with investments with rates of return that fluctuate. Stock prices, in particular, fluctuate by the minute. If you buy a stock that has on average been increasing in price, but that has also been varying a lot around that average trend, how can you project the spread of possible returns (including possible negative returns) into the future?

Recall our exponential growth model from Chapter 10 that projects n, the amount of money (or other exponentially growing stuff, like population biomass) t time units into the future. The model equation was

$$n = me^{rt},$$

where m is the initial of money invested and r is the annual interest or growth rate with the interest being compounded continuously or over short time intervals such as daily or hourly. On the logarithmic scale, the growth of the money was a linear function of time:

$$\log(n) = \log(m) + rt.$$

We take this as a base model for our investment, but as it is, it is not enough. The exponential model is just a model of the trend. We need to add in the random fluctuations that make stock prices so volatile. Let us look at the trend model as a short-term model of change. Suppose we let t be a small time interval, such as a minute. Suppose we let x denote the log-amount of money at the beginning of the interval: $x = \log(m)$. Then according to just the trend model, the new log-amount of money x_{new} after the small time interval t would be

$$x_{new} = x + r(\Delta t).$$

Another way of writing this is that the change $x = x_{new} - x$ in the log-amount of money over the small time interval is just r times t:

$$\Delta x = r(\Delta t).$$

Here is where we can work with adding in some random component to investment growth. Our problem of the fluctuating rate of return can be restated as: "the amount of change in our money from minute to minute fluctuates randomly." The restatement suggests adding a small amount of randomness to the amount of change above, then calculating the new amount of money, then using the new amount of money to project the money for another small time interval Δt into the future, and so on. It will take a loop with many cycles to project very far into the future, but we have the computing capacity for the job. Also, adding randomness on the logarithmic scale is like multiplying by randomness on the original scale, which is consistent with the multiplicative nature of the growth of investments.

We can think of calculating the change each time as

$$\Delta x = r(\Delta t) + W,$$

where W is a random variable that is generated anew for each small time interval. An initial model might be to assume W has a normal distribution that is centered around 0 (i.e., $\mu = 0$); departures from our long-term rate of change, when examined minute by minute, are likely negative as well as positive. The spread constant σ would be something quite small—we think of it as how large a typical departure from the growth amount $r(\Delta t)$ would be during a small time interval.

Let us get organized and summarize the computational steps needed.

Step 0: Initialization. Set numerical values for r, Δt, and the initial investment amount that will be used in the simulation. Ultimately, the value of r we use would be based on the recent past performance of the investment, but once the script is written and working, we can alter r to project different hypothetical scenarios. We will need to do some thinking about the random "noise" we are going to add to each incremental change and set the values of any constants used in the noise at this step. Finally, we will need to set up some vectors in which we need to store results for plotting and analysis.

We will begin with an attractive-looking investment that has been averaging around 6% per year. So, $r = .06$. We have been thinking of our little time interval Δt as 1 minute. Now, r applies to a whole stock trading year, say, 260 8-hour trading days. There are $260 \times 8 \times 60 = 124{,}800$ trading minutes in 1 trading year, and so 1 minute is 1/124,800th of a trading year. So, $\Delta t = 1/124{,}800$.

We are assuming that our "noise" random variable has a centering constant of $\mu = 0$. We might think of the spread constant as some fraction of r, scaled by the small time interval Δt. For technical reasons, in these types of noisy investment models (which are really used in investment banks), the frequently used time scaling term is $\sqrt{\Delta t}$. So, if we think of the typical noise fluctuation on an annual basis as 20% of r, then our spread constant is $\sigma = .2r\sqrt{\Delta t}$.

Our initial investment is \$1000. Suppose we want to project the price of our stock for 3 months (or, let us say 60 trading days). This will total $60 \times 8 \times 60 = 28{,}800$ trading minutes. We will need a vector with these many elements (plus 1 additional element for the initial size) to store all the minute-by-minute investment prices. We will also need another vector of similar size to store the accumulated times.

Step 1: The simulation. Generate a vector of 28,800 random noise variables. Project the investment prices inside a for-loop, with each time through the loop representing 1 trading minute. Each time through the loop, the change in log-price of the investment gets punched by a new random noise variable. Each time through the loop, accumulated time gets updated.

Step 2: Exponentiate (take antilog) all the log-prices so that they represent prices. Plot prices as a function of time.

Here is one script that performs the above simulation:

```
#=============================================
# R script to simulate investment with randomly
# fluctuating prices.
#=============================================

# Step 0.  Initialization.
days=60                 # Time horizon of simulation (trading
                        #   days).
r=.06                   # Annual average return of 6%.
dt=1/124800             # 124800 minutes in 260 8-hr trading days
                        #   per year.
sig=.2*r*sqrt(dt)       # Spread constant for noise.
mnts=days*8*60          # Number of minutes in 60 trading days.
x=numeric(mnts+1)       # Vector to contain the log-prices.
t=x;                    # Vector to contain the accumulated times.
x[1]=log(1000)          # Initial investment amount.
t[1]=0                  # Initial time.

# Step 1.  Simulation.
w=rnorm(mnts,0,sig)     # Generate vector of normal noises outside
                        #   the loop (more efficient)
for (i in 1:mnts) {
   dx=r*dt+w[i]         # Change in log-price during one minute.
   x[i+1]=x[i]+dx       # New price after one minute.
   t[i+1]=t[i]+dt       # New accumulated time after one minute.
}

# Step 2.  Plotting.
n=exp(x)                # Change log-prices to prices.
plot(t,n,type="l")      # Plot prices vs time.
```

The script produced Figure 14.6.

Run the script several times. What happens to the plot? It is considerably different each time! After buying this stock, we are in for a thrilling roller coaster ride, in which we never know what is coming next.

It seems, in fact, that just one plot does not really give us enough information. We would really like to know the range and distribution of the possible

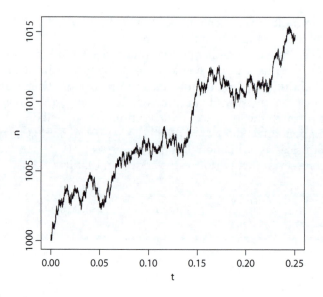

FIGURE 14.6
Simulation of a randomly fluctuating stock price using the noisy exponential growth model.
Time unit is 1 trading year.

stock prices after 60 trading days in order to estimate the risks and possible rates of return if we sell at the end of the 60 days. We need several such plots. We need 1000 such plots!

Well, 1000 plots might be a bit unwieldy. However, we could embed our simulation in an outside loop and repeat it 1000 times. We could easily store 1000 ending prices, one from each simulation, in a vector. We could then draw a histogram!

The previous script only needs a little modification. We need a vector to store the final stock prices. We need to build an outside for-loop, initializing the t and x vectors each time, sort of like starting a DVD over again 1000 times. We need to delete the price-time plot and draw a histogram instead:

```
#===========================================
# R script to simulate investment with randomly
# fluctuating prices many times and draw histogram
# of final prices.
#===========================================

# Step 0.  Initialization.
days=60             # Time horizon of simulation (trading days).
sim.num=1000        # Number of simulations.
r=.06               # Annual average return of 6%.
dt=1/124800         # 124800 minutes in 260 8-hr trading days
                    #   per year.
sig=.2*r*sqrt(dt)   # Spread constant for noise.
```

```
mnts=days*8*60        # Number of minutes in 60 trading days.
x=numeric(mnts+1)     # Vector to contain the log-prices.
t=x;                  # Vector to contain the accumulated times.
log.prices=numeric(sim.num)

# Step 1.  Simulation.

w=matrix(rnorm(mnts*sim.num,0,sig),sim.num,mnts)
for (h in 1:sim.num) {
  x[1]=log(1000)        # Initial investment amount.
  t[1]=0                # Initial time.

  for (i in 1:mnts) {
    dx=r*dt+w[h,i]      # Change in log-price during one minute.
    x[i+1]=x[i]+dx      # New price after one minute.
    t[i+1]=t[i]+dt      # New accumulated time after one minute.
  }
  log.prices[h]=x[mnts+1]  # Last log-price in x
}

# Step 2.  Plotting.

prices=exp(log.prices)    # Change log-prices to prices.
hist(prices,freq=FALSE)   # Histogram.
```

Run the script. It takes a while, be patient! The script produces a histogram approximately like Figure 14.7. Small differences will appear each time the script is run, but the main pattern will be the same. The area of the rectangle

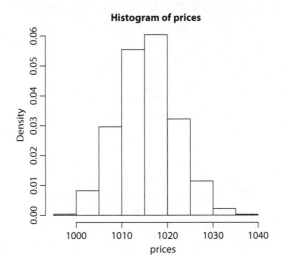

FIGURE 14.7
Histogram of closing prices after 60 trading days from 1000 simulations of a randomly fluctuating stock price using the noisy exponential growth model. Initial price was $1000. Time unit is 1 trading year.

over a price interval is the estimated probability that the stock price will be in that interval after 60 trading days. One can see from Figure 14.7 that there is some small but real risk that the price after 60 days will be less than the initial price of $1000. There is a moderately large chance that the price after 60 days will be greater than $1025.

WHAT WE LEARNED

1. A random variable is a quantity that varies randomly.
2. The Law of Large Numbers states that if a random process is repeated many times, the long-run proportion of times that a particular outcome occurs becomes more and more stable. The long-run proportion stabilizes around (converge to) a quantity that is called the probability of that outcome.
3. A binomial random process has the following characteristics. (1) A collection of n trials takes place, with each trial having only two possible outcomes: one traditionally labeled "success" and the other one labeled "failure." (2) For each trial, the probability of success, denoted p, is the same. (3) The trials are independent, that is, the outcome of one trial does not affect the outcome probability for any other trial. The count of the number of successes is a random variable with a binomial distribution and can take possible values 0, 1, 2, 3, ..., n.
4. R functions for calculating and simulating aspects of the binomial distribution are dbinom(k,n,p) (probability of k successes), pbinom(m,n,p) (sum of probabilities from 0 to m), and rbinom(size,n,p) (generates vector of binomial random variables n with length size).

Examples:

```
> n=10
> p=.4
> k=0:10
> dbinom(k,n,p)
 [1]   0.0060466176 0.0403107840 0.1209323520
 [4]   0.2149908480 0.2508226560 0.2006581248
 [7]   0.1114767360 0.0424673280 0.0106168320
[10]   0.0015728640 0.0001048576
> m=8
> pbinom(m,n,p)
[1] 0.9983223
> size=5
> rbinom(size,n,p)
[1] 2 5 3 7 4
```

5. A uniform distribution is a continuous distribution on the interval from 0 to 1. The density curve for a uniform distribution is a flat line of height 1 over the interval 0 to 1. The probability that a uniform random variable U takes a value between a and b is the area under the flat line between a and b and is given by

$$P(a \leq U \leq b) = b - a,$$

where $0 \leq a \leq b \leq 1$.

6. The R function `runif(size)` simulates a vector of length `size` of uniform random variables on the interval between 0 and 1.

7. The Central Limit Theorem states that sums of random variables tend to have distributions that are adequately approximated with a normal distribution. The normal distribution has a bell-shaped density curve characterized by a centering constant μ and a spread constant σ.

8. R functions that calculate aspects of the normal distribution are `dnorm(y,mu,sigma)` (calculates the height of the normal density curve over y), `pnorm(y,mu,sigma)` (calculates the area under the normal density curve to the left of y), and `rnorm(size,mu,sigma)` (simulates a vector of normal random variables of length `size`).

Computational Challenges

14.1. For the fluctuating stock price model, what happens to the chance that the 60-day closing price will be below 1000 and the chance that it will be above 1020 if the "volatility" σ is increased to, say, $.6r\sqrt{t}$?

14.2. Otto claims to be psychic and to be able to predict the suit of the top card of a shuffled deck of cards. A skeptical friend challenges Otto to demonstrate his abilities. The friend presents Otto with a shuffled deck and Otto states a prediction. The friend reveals the card and records whether Otto was correct or not. The friend and Otto repeat the process 100 times: shuffling, guessing, and recording. Otto gets 30 correct. If Otto is not psychic and is just guessing, the chance of success on any trial is .25. Otto claims that 30/100 is bigger than .25 and so he has demonstrated evidence for his claim of being psychic. What is the chance that Otto would get such an extreme result, that is, what is the chance that he would get 30 *or more successes* if he was just guessing? In light of this calculation, how convincing is his evidence?

14.3. You are going to build your dream house on a parcel of land. The parcel is on a 500-year flood plain. One way to understand the probability meaning of a 500-year flood plain is as follows. Think of a bowl of 500 marbles. The bowl has 499 red marbles and 1 blue marble in it. Each year, nature draws a marble from bowl, notes the color, and puts the marble back in the bowl, where it gets thoroughly mixed back in before the next year. If the marble is red, there is no flood on the parcel that year. If the marble is blue, there is a flood on the parcel that year. In the next 10 years, the number of floods on the parcel could be 0, or 1, or 2, or 3, or any integer up to and including 10 (a flood each year). Calculate the probability for each possible number of floods in the next 10 years (the probability of 0, of 1, ..., of 10). What is the probability of one or more floods in the next 10 years? Are you still going to build that house?

Afternotes

1. A stock future option is the right to purchase or sell shares of a stock at a fixed price at a certain time in the future. The Black–Scholes model (Black and Scholes 1973) is a famous mathematical model in economics and finance that calculates the price of an option over time. The Black–Scholes model uses as its basis a stock price fluctuation model much like the one explored in the "Real-World Example" section. The model has been widely adopted by investment firms. For this work, Black was awarded the Nobel Prize in economics in 1997 (Scholes died in 1995 and was therefore ineligible for the prize).

2. The stock price fluctuation model is a noisy model of exponential growth. In that sense, it can serve as a model of the growth of an endangered species population in the presence of fluctuating environmental conditions. The model can be used to estimate and simulate the risk that a species will reach extremely low levels within a given time horizon (Dennis et al. 1991).

3. The world runs on the Law of Large Numbers. Insurance corporations and gambling casinos, for instance, count on long-run stable proportions of outcomes of random events in order to have reliable streams of incomes.

References

Black, F., and M. Scholes. 1973. The pricing of options and corporate liabilities. *Journal of Political Economy* 81:637–654.

Dennis, B., J. M. Scott, and P. L. Munholland. 1991. Estimation of growth and extinction parameters for endangered species. *Ecological Monographs* 61:115–143.

15

Fitting Models to Data

Fitting a Quadratic Model

In Chapter 8, we studied a model of sustainable harvesting of renewable resources. The model was a quadratic curve relating the sustainable yield Y to the amount of harvest effort E:

$$Y = hE - gE^2.$$

Here, h and g are constants with values that will be different from resource to resource and situation to situation. The constants ultimately depend on the rate of growth of the biological resource, that is, the rate at which the resource is replenished. However, the data we had in hand for the example, shrimp in the Gulf of Mexico (Table 8.1), only provided information about Y under varying values of E. No information about the biology or population growth rates of shrimp were provided toward determining the constants h and g.

We will use a **curve-fitting method** to determine the values of h and g that yield the quadratic model with the best description of the data. By "best" we mean the quadratic that optimizes some sort of measure of prediction quality. Recall that we used the least squares method in Chapter 12 for finding the slope and the intercept of a line for prediction, when the relationship between variables appeared linear. Least squares actually is a common criterion used in science for fitting curves as well as lines. Extending the least squares concept to a quadratic is easy, and calculating the coefficients is even easier.

Let us look at the fitting problem for a general quadratic and then apply the methods to the sustainable yield model and the shrimp data. We suppose that there is a variable y that we want to predict and a variable x to be used for predicting y. We suppose that observations on x and y exist in the form of n ordered pairs $(x_1, y_1), (x_2, y_2), ..., (x_n, y_n)$ that appear to suggest a quadratic relationship when plotted in a scatterplot (such as Figure 8.6). We hypothesize from scientific grounds that a quadratic model of the form

$$y_{predicted} = b_1 + b_2 x + b_3 x^2$$

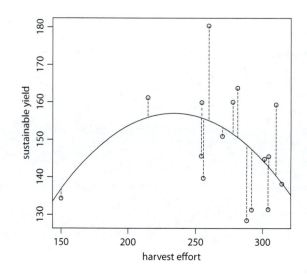

FIGURE 15.1
The dashed lines are the prediction errors or vertical departures of the data (circles) from the parabola (solid curve). The sum of squared prediction errors is minimized by the least squares curve.

will adequately describe the data, where b_1, b_2, and b_3 are constants with values to be determined by curve fitting. We use the notation b_1, b_2, and b_3 for the coefficients instead of c, b, and a as in Chapter 8, in anticipation that the coefficients will be elements of a vector.

We define the prediction error for the ith observation to be $y_i - (b_1 + b_2 x_i + b_3 x_i^2)$, that is, the vertical distance (with a plus or minus sign) between the value of y_i and a given prediction curve calculated with the value of x_i (Figure 15.1). For a given curve (i.e., a quadratic with given values of b_1, b_2, and b_3) and a given set of observations, the sum of squared prediction errors is

$$\left(y_1 - b_1 - b_2 x_1 - b_3 x_1^2\right)^2 + \left(y_2 - b_1 - b_2 x_2 - b_3 x_2^2\right)^2 + \ldots$$
$$+ \left(y_n - b_1 - b_2 x_n - b_3 x_n^2\right)^2.$$

The least squares estimates of b_1, b_2, and b_3 are the values that would produce the minimum sum of squared errors. In Chapter 12, to calculate the least squares estimates of the slope and the intercept of a line, we used Legendre's result, that the least squares estimates are the solution to a system of linear equations. Legendre's original result extends to the coefficients of a quadratic. Matrix mathematics, and of course computers, were invented well after Legendre's time, and we will take full advantage of both. First, we build a matrix **X** of the predictor variable data. Each line of the matrix will represent the predictor data for one observation. The matrix will have a column corresponding to each of the three terms b_1, $b_2 x$, and $b_3 x^2$ in the prediction equation. For the ith row, the first element will be a 1, the second element will be x_i, and the third element will be x_i^2:

$$\mathbf{X} = \begin{bmatrix} 1 & x_1 & x_1^2 \\ 1 & x_2 & x_2^2 \\ \vdots & \vdots & \vdots \\ 1 & x_n & x_n^2 \end{bmatrix}.$$

The data for the variable y are then arranged into a column vector:

$$\mathbf{y} = \begin{bmatrix} y_1 \\ y_2 \\ \vdots \\ y_n \end{bmatrix}.$$

The column vector of least squares estimates according to Legendre's result will be the solution to a system of linear equations given by

$$(\mathbf{X'X})\mathbf{b} = \mathbf{X'y},$$

which is a matrix expression of three equations in three unknowns. Remember that $\mathbf{X'}$ denotes the transpose of the matrix $\mathbf{X}$ (the matrix $\mathbf{X}$ with columns turned into rows). The solution $\hat{\mathbf{b}}$ if it exists is calculated with the inverse of the matrix $(\mathbf{X'X})$:

$$\hat{\mathbf{b}} = \begin{bmatrix} \hat{b}_1 \\ \hat{b}_2 \\ \hat{b}_3 \end{bmatrix} = (\mathbf{X'X})^{-1}\mathbf{X'y}.$$

The sustainable yield model is in the form

$$y_{\text{predicted}} = b_1 x + b_2 x^2,$$

which is a quadratic equation that intersects the vertical axis at 0. Such a quadratic is fitted to data by omitting the column of 1s from the $\mathbf{X}$ matrix.

Recall the script from Chapter 8 that plotted the shrimp sustainable yield data and the fitted quadratic model. Let us modify it to include the actual fitting of the quadratic model:

```
#================================================================
# R script to draw a scatterplot of sustainable yield vs.
# harvesting effort for shrimp data from the Gulf of Mexico,
# with parabolic yield-effort curve fitted and superimposed.
#================================================================
```

```
#-----------------------------------------------------------
# 1.  Read and plot the data.
#  (a data file named shrimp_yield_effort_data.txt is assumed
#  to be in the working directory of R)
#-----------------------------------------------------------
df=read.table("shrimp_yield_effort_data.txt",header=TRUE)
attach(df)
plot(effort,yield,type="p",xlab="harvest effort",
  ylab="sustainable yield")
#-----------------------------------------------------------
# 2.  Fit parabolic yield-effort curve with least squares.
#-----------------------------------------------------------
X=cbind(effort,effort^2)  # First column of matrix X is effort.
                          # Second column is effort squared.
b=solve(t(X)%*%X,t(X)%*%yield)  # Least squares solution.
h=b[1,]           # Coefficient of effort.
g=b[2,]           # Coefficient of effort squared.
detach(df)

#-----------------------------------------------------------
# 3.  Overlay fitted quadratic model on scatterplot.
#-----------------------------------------------------------
elo=100               # Low value of effort for calculating
                      # quadratic.
ehi=350               # High value of effort.
eff=elo+(0:100)*(ehi-elo)/100  # Range of effort values
sy=h*eff+g*eff^2      # Sustainable yield calculated for range of
                      #  effort values.
points(eff,sy,type="l")  # Add the quadratic model to the plot.
```

Just a few lines in part 2 of the script were altered for calculating the values of g and h instead of just assigning them. Run the script and Figure 8.6 is reproduced. We remark again that the shrimp data are so variable that a quadratic shape is a barely recognizable pattern. Improved models might result from adding one or more additional predictor variables such as variables related to prevailing ocean conditions or economic conditions that year.

Multiple Predictor Variables

The x^2 term in the predictive equation earlier plays the role of a second predictor variable. More terms with different predictor variables can be added to such models, provided the number of observations is sufficient to support the estimation of the coefficients.

As an example, Computational Challenge 13.1 listed 8 years of data collected by the U.S. Bureau of Land Management on the pronghorn (*Antilocapra americana*) population in the Thunder Basin National Grassland in Wyoming. The area surrounding the reserve is experiencing environmental degradation from fossil fuel extraction and intensive livestock grazing. The main interest of the wildlife managers is in the success of reproduction, as measured by the number of fawns. Reproductive success in wildlife is often a key indicator of habitat quality and environmental conditions. The pronghorn is an important game species, popular among hunters due to the challenge of hunting this speedy and alert animal.

We can attempt to build a prediction equation using the fawn count y as the variable to be predicted and the three variables (pronghorn population size u, annual precipitation v, and winter severity index w) as predictors. The prediction equation would have the form

$$y_{predicted} = b_1 + b_2u + b_3v + b_4w.$$

The values b_1, b_2, b_3, and b_4 are unknown constants to be determined with the least squares method. Having four constants to estimate with only eight observations is pushing the limits of validity of least squares, but the resulting model can at least serve the wildlife biologists as a hypothesis for testing as additional data become available. The X, y, and b matrices for fitting the model are as follows:

$$X = \begin{bmatrix} 1 & 920 & 13.2 & 2 \\ 1 & 870 & 11.5 & 3 \\ \vdots & \vdots & \vdots & \vdots \\ 1 & 790 & 11.2 & 3 \end{bmatrix}, \quad y = \begin{bmatrix} 290 \\ 240 \\ \vdots \\ 210 \end{bmatrix}, \quad b = \begin{bmatrix} b_1 \\ b_2 \\ b_3 \\ b_4 \end{bmatrix}.$$

The least squares solution equation is the same as before. Legendre's result covers any number of predictor variables. Be aware, however, that Legendre's result does not say anything about the quality of the resulting model; it is merely a prescription for fitting a model using least squares.

A slight alteration of the least squares script from Chapter 12 produces the estimates of b_1, b_2, b_3, and b_4. With more than one predictor variable, it is difficult to draw model and data in their full dimensions. However, we can plot the fawn counts through time and plot the model predictions for those times, to get an idea about whether the model describes the data well or not:

```
#==================================================================
# R script to calculate least squares estimates for pronghorn
# data.
# Variable for prediction is y, spring fawn count.
# Predictor variables are size of adult pronghorn population (u),
# annual inches precipitation (v), winter severity index (w;
# scale of 1:mild-5:severe).
```

```
# Script produces a time plot of the data along with overlayed
# predictions.
#================================================================

#-----------------------------------------------------------
# Input the data.
#-----------------------------------------------------------
Pronghorn=read.table("pronghorn.txt",header=TRUE) # Change file
                                                  # name if
                                                  # necessary.
attach(Pronghorn)

#-----------------------------------------------------------
# Calculate least squares intercept and slope.
#-----------------------------------------------------------
n=length(y)              # Number of observations is n.
X=matrix(1,n,4)          # Form the X matrix:  col 1 has 1's,
X[,2]=u                  # cols 2-4 have predictor
                         # variables.
X[,3]=v                  #        ---
X[,4]=w                  #        ---
b=solve(t(X)%*%X,t(X)%*%y)    # Least squares estimates in b;
                              # t() is transpose function.
                              # Alternatively can use
                              # b=solve(t(X)%*%X)%*%t(X)%*%y.

#-----------------------------------------------------------
# Draw a time plot of data, with superimposed least
# squares predictions.
#-----------------------------------------------------------
plot((1:8),y,type="o",pch=1,xlab="time (yr)",
  ylab="spring fawn count")    # Time plot of data.

ypredict=X%*%b  # Calculate predicted y values at values of #
# predictors.

points((1:8),ypredict,type="p",pch=2)   # Plot predicted values
                                        # with points.

#-----------------------------------------------------------
# Print the least squares estimates to the console.
#-----------------------------------------------------------
"least squares intercept and coefficients: "
b
detach(Pronghorn)
```

Run the script. Figure 15.2 results. The predictions are close to the data, but that could be an artifact of having few data and many predictors.

The problem of evaluating the quality of fitted models is an important but advanced topic. Fitting models to data and evaluating the models is called

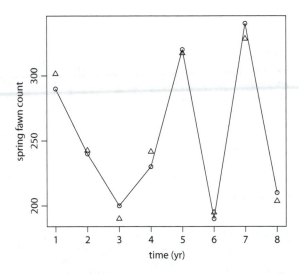

FIGURE 15.2
Circles: Spring fawn counts of pronghorn (*Antilocapra americana*) population in the Thunder Basin National Grassland in Wyoming. Triangles: Predicted fawn counts from a model with pronghorn population size, annual precipitation, and winter severity index as predictor variables.

regression analysis, and entire college courses are devoted to its study. A model with just one predictor variable is the topic of **simple linear regression**, and a model with multiple predictor variables is the topic of **multiple regression**. One key idea that emerges in the study of regression is that adding more predictor variables involves a trade-off; while more predictor variables can make a model fit the existing data better, predictions of *new* data are badly degraded by too many predictor variables! In any regression application, getting the balance right is a scientific challenge. And, it is always a good idea to regard fitted models not as settled knowledge but rather as hypotheses to be tested against new data. In Computational Challenge 15.2 at the end of the chapter, you will be invited to fit several different models to the pronghorn data and calculate a numerical "index" of model quality that is currently popular in the sciences.

Nonlinear Statistical Models

The models like

$$y_{\text{predicted}} = b_1 + b_2 x + b_3 x^2$$

and

$$y_{\text{predicted}} = b_1 + b_2 u + b_3 v + b_4 w$$

that we have worked with so far in this chapter are called **linear statistical models**. The word "linear" in this term does not refer to the $y_{predicted}$ versus predictor relationship—the quadratic model discussed earlier, for instance, is a curved function of x, not a linear function. Rather, "linear" refers to the unknown coefficients b_i; they appear linearly in the model equations, that is, no terms like b_i^2 or e^{b_i} appear in the equations.

Various mathematical models in science have constants that enter nonlinearly. Such models might arise from mechanisms hypothesized and derived for particular scientific applications. For instance, recall the data from Chapter 1 on how the rate at which a wolf kills moose varies with the moose supply. On the scatterplot of the data (Figure 1.2), we overlaid an equation that has the general form

$$y = \frac{b_1 x}{b_2 + x},$$

where y is the number of moose killed per wolf in a period of time, and x is the density of moose. Here, y is not a linear function of the constants b_1 and b_2. Also, the constants b_1 and b_2 vary among different predators and different prey and are usually estimated by fitting the function to data.

The kill rate model (called the functional response model by ecologists) arises from scientific considerations. It is a model of "diminishing returns," in which adding more moose at high moose density produces less additional moose kill than adding the same amount of moose at low moose densities. The mechanism in such phenomena is often some sort of bottleneck in the process by which the quantity on the vertical axis, in this case, dead moose, is produced. A typical wolf reaches a biological processing capacity—the wolf's digestive system can only handle so much meat at a time.

Similar bottlenecks occur in other systems. For instance, in an enzyme-catalyzed chemical reaction, an enzyme molecule temporarily binds with a substrate molecule, which helps to transform the substrate into a product molecule. The enzyme then releases the product and continues on to transform another substrate molecule into product. In such systems, there is only a finite amount or concentration of enzyme molecules. When almost all the enzyme molecules are "busy," adding more substrate does not appreciably increase the rate of product formation (just like adding more people to the line of people waiting for the next available clerk does not increase how fast people are serviced). Indeed, the kill rate equation above is identical to the model that biochemists use to describe how the rate of product formation (y) depends on the substrate concentration (x) in an enzyme-catalyzed reaction. In the accompanying box, a derivation of the functional response equation is given using just algebra to give an idea of the underlying reasons why biologists use this particular equation as a model.

Whether for wolves or enzymes, the rate equation frequently must be fitted to data, that is, the values of b_1 and b_2 that provide the best rate equation for describing observations on y and x must be determined. As was the case with linear statistical models, least squares is a reasonable criterion for fitting. Least squares would select the values of b_1 and b_2 that minimize

$$\left(y_1 - \frac{b_1 x_1}{b_2 + x_1}\right)^2 + \left(y_2 - \frac{b_1 x_2}{b_2 + x_2}\right)^2 + \ldots + \left(y_n - \frac{b_1 x_n}{b_2 + x_n}\right)^2.$$

Here, $y_1, y_2, \ldots, y_n$ are values of y that were measured experimentally or observationally at the different values $x_1, x_2, \ldots, x_n$ of x.

The problem with least squares here is computational. The kill rate equation is a **nonlinear statistical model**, in which one or more of the constants (in this case, b_2) enter the equation in a nonlinear fashion. The computational problem stems from the fact that Legendre's result only applies to linear statistical models. We do not have any formula available for calculating least squares estimates for nonlinear statistical models.

Fortunately, we have computers, and we have R. R contains a built-in function for **numerical minimization** of a sum of squares. The function is called nls(), for nonlinear least squares.

To use the function, you need to have initial ballpark guesses about the values of the constants to be estimated. The function takes the initial values and calculates a correction that will make the sum of squares lower. The correction calculation is then repeated for the new values of the constants, lowering the sum of squares even more. The function **iterates** (repeats) the correction calculation until the sum of squares can be lowered no more. Like a fly in a sugar bowl, the nonlinear least squares calculations crawl down the sum of squares surface until the only direction remaining is up.

The initial guesses ideally should be in the vicinity of the least squares estimates. The sum of squares surface in nonlinear models can sometimes have multiple low points (like two valleys separated by a ridge), and the desired estimates then correspond to the lowest valley. Or, the iterative calculations might start spinning off into outer space if, for instance, the initial value of a positive constant was taken to be negative. To get good initial guesses, it is helpful to interpret what the constants represent quantitatively in the model. For example, in the kill rate equation, the constant b_1 is the upper limit (asymptote) for the curve, and b_2 is the value of x where the curve is halfway up to the asymptote. Initial values for these constants can be obtained by inspecting a plot of the data. Often, rerunning the calculations from different initial values helps the analyst to explore the least squares surface and warn of the presence of "multiple local minima."

That there are lots of things to be cautious about in fitting nonlinear statistical models detracts only slightly from the fact that nls() in R is really easy to use. What follows is a small script for fitting the kill rate equation and drawing a plot of data overlaid with fitted model. We will use the wolf–moose data from Chapters 1 and 2. Our first script from Chapter 2 to plot the data and fitted model together can serve as a starting point, but by now in this book we can work with the data in a file and finish with more descriptive axis labels on the graph:

```
#================================================================
# R script to calculate nonlinear least squares estimates for
# parameters b1 (maximum feeding rate) and b2 (half saturation
# constant) in the rectangular hyperbolic equation for feeding
# rate (Holling type 2 functional response, Monod nutrient uptake
# rate, Michaelis-Menten enzyme equation).  The equation is defined
# by
#                b1*prey
#   rate = ----------     .
#             b2 + prey
#
# Here 0<b1, 0<b2, and "prey" is the density, concentration, or
# abundance of prey, substrate, or nutrient.
#
# Data in example are moose density (# per 1000 km^2) and number
# killed per wolf in 100 d from:
# Messier, F.  1994.  Ungulate Population Models with Predation: A
# Case Study with the North American Moose.  Ecology 75:478-488.
#================================================================

#----------------------------------------------------------------
# 1.  Input the data.  Text file of data has two columns, one
# labeled
# "prey" and one labeled "rate".
#----------------------------------------------------------------
Wolf.Moose=read.table("wolf_moose.txt",header=TRUE) # Change
                                                    # file name
                                                    # if
                                                    # necessary.
attach(Wolf.Moose)
#----------------------------------------------------------------
# 2.  Calculate initial values using a linearization
# transform. The transform allows initial values
# to be calculated with a multiple regression
# without intercept:
#                rate*prey = b1*prey - b2*rate.
#----------------------------------------------------------------
yy=rate*prey;
xx=cbind(prey,rate);
bb=solve(t(xx)%*%xx,t(xx)%*%yy)
b1.0=bb[1];
b2.0=-bb[2];

#----------------------------------------------------------------
# 3.  Use nls() function to minimize sum of squares with an
# iterative numerical method.  The object "curve.fit" will be a list
# of results from the calculations.
#----------------------------------------------------------------
```

```
curve.fit=nls(rate~b1*prey/(b2+prey),data=Wolf.Moose,
  start=list(b1=b1.0,b2=b2.0))

#------------------------------------------------------------
# 4.  Print the results of the calculations.  Assign values to b1
# and b2 for plotting the fitted model.
#------------------------------------------------------------
summary(curve.fit)
b=coef(curve.fit)
b1=b[1]
b2=b[2]


#------------------------------------------------------------
# 5.  Calculate the fitted rate curve for a range of values of
# prey, and store the values in "prey.vals".
# Range is from 0 to slightly beyond max(prey).  Change range if
# desired.  Values of fitted rate curve are in "fitted.curve".
#------------------------------------------------------------
prey.vals=(0:100)*1.01*max(prey)/100;
fitted.curve=prey.vals*b1/(b2+prey.vals);


#------------------------------------------------------------
# 6.  Plot the data in a scatterplot.  Overlay the fitted rate
# equation.
#------------------------------------------------------------
plot(prey,rate,ylim=c(0,4),xlim=c(0,2.5));
points(prey.vals,fitted.curve,type="l");


#------------------------------------------------------------
# 7.  Tidy up a bit.
#------------------------------------------------------------
detach(Wolf.Moose)
```

Let us review some of the features of this script. In part 1 of the script, a data file is read and the data frame Wolf.Moose is created. The column names in the data file and subsequent variables in the data frame are prey and rate (Table 15.1).

In part 2, initial values of the constants b_1 and b_2 are calculated with linear least squares. The predator–prey rate equation,

$$y = \frac{b_1 x}{b_2 + x},$$

can be algebraically rearranged as

$$yx = b_1 x - b_2 y$$

TABLE 15.1

Prey: Number of Moose per 1000 Square Kilometers
Rate: Number of Moose Killed per Wolf per 100 Days

prey	rate
0.17	0.37
0.23	0.47
0.23	1.90
0.26	2.04
0.37	1.12
0.42	1.74
0.66	2.78
0.80	1.85
1.11	1.88
1.30	1.96
1.37	1.80
1.41	2.44
1.73	2.81
2.49	3.75

Source: Messier, F., *Ecology*, 75, 478–488, 1994.

(try this). The rearrangement suggests doing a linear least squares fit using yx as the variable to be predicted and x and y as predictors, in a model without an intercept constant. This is how the initial values (named b1.0 and b2.0 in the script) of the constants are obtained. Many nonlinear equations commonly fitted to data can be similarly rearranged into a linear form of some sort. The predator–prey rate equation actually has several different linear forms; before computers were in widespread use, biologists used the linearizations as the way of estimating the constants because the calculations were within the reach of a mechanical or electric calculator. Unfortunately, the linear forms for many models typically do not yield the best values of the constants for predicting y. The linear forms, however, can provide ballpark values of the constants that serve fine as initial values in the nonlinear least squares calculations. Our script is thereby automated somewhat, in that it could be used for enzyme or predator–prey data sets of all sorts, without the user having to puzzle out good initial values for each application.

Part 3 is the heart of the script. The nls() function is invoked to calculate least squares estimates for the rate equation. In the function, the first argument is rate~b1*prey/(b2+prey). The vectors rate and prey are in the workspace, while b1 and b2 are the constants to be estimated. On the left of the squiggle is the variable to be predicted, while on the right is a R expression giving the form of the equation to be fitted. The second argument is start=list(b1=b1.0,b2=b2.0). The argument uses the list() function of R, which creates a list of R objects, in this case, statements assigning initial values to b1 and b2. The nls() function when invoked creates a list of output, which in the script is named curve.fit.

In part 4, the script uses the summary() function to print the results in curve.fit to the console. A vector named b containing the least squares estimates for the constants is extracted from curve.fit with the coef() function (the coef() function is a special function that works with all the modeling functions in R). The constants b1 and b2 are respectively the first and the second elements of b. The order of the constants in b corresponds to the order of the initial values specified in the nls() function. The values of b1 and b2 appear at the console along with additional statistical information. Explaining the additional statistical information is beyond the scope of this book, but such an explanation is a standard part of any college statistics course.

Part 5 calculates the fitted predator–prey rate curve for 100 different values of x. The values range from 0 to $1.01 \times$ the largest x value in the data, so that the curve will extend slightly beyond the rightmost data point on the graph.

Part 6 plots the data with plot() and then overlays the fitted model using points().

Part 7 detaches the data frame in case additional analyses using other data are to be performed.

Run the script, and Figure 15.3 will be reproduced, along with the results printed at the console. At the all-you-can-eat moose buffet, the average wolf maxes out at roughly 3.4 moose eaten per 100 days.

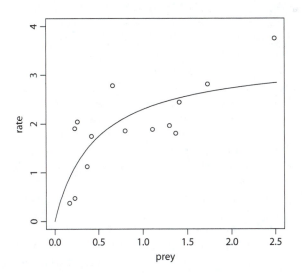

FIGURE 15.3
Data and model depicting the relationship between the wolf rate of feeding on moose and the density of moose. Circles: number of moose killed per wolf per 100 days (vertical axis) with the number of moose per 1000 square kilometers. Solid line: Holling type II functional response equation from predator–prey theory (Gotelli 2008) fitted to the data. (Data and fitted equation from Messier, F., *Ecology*, 75, 478–488, 1994.)

Final Remarks

REMEMBER: Mathematical models of real-world phenomena are hypotheses. We derive them on scientific grounds, try them out on data, and test them with experiments and new data. Our reliance on a model starts out tentatively and grows gradually as the model proves itself more and more to be reliable.

DERIVING THE FUNCTIONAL RESPONSE EQUATION

Biologists take fairly seriously the functional response equation, given by

$$y = \frac{b_1 x}{b_2 + x},$$

as a model of the feeding rate of a predator, of the rate substrate depletion in an enzyme reaction, and of the rate of nutrient uptake by bacteria. The function describes many different data sets in many different situations (see the Computational Challenges). Moreover, deriving the equation based on biological mechanisms is straightforward. We will go through one derivation here, which captures the basic ideas.

We will cast the derivation in terms of an enzyme-catalyzed reaction in biochemistry. The quantities and aspects of the derivation have direct analogies in other systems, which will be pointed out at the end.

Each enzyme molecule binds with a substrate molecule temporarily and catalyzes (facilitates) a chemical reaction that turns the substrate molecule into a product molecule. For instance, the enzyme amylase in your saliva breaks starch (a chain of glucose sugar molecules) into individual glucose molecules. The enzyme molecule is not changed by the reaction but rather lives on to bind with another substrate molecule.

Suppose we denote by E the total concentration of enzyme molecules. For the time period of the experiment or observation, E is assumed not to change and is going to be the ultimate cause of the bottleneck in the reaction rate. We denote by B the concentration of bound enzyme molecules, that is, the molecules that are "busy" transforming substrate molecules into product. Thus, $E - B$ is the concentration of enzyme molecules that are free to "service" substrate molecules.

We denote by S the substrate concentration. The principal quantity we are after is the change in S in a small unit of time, divided by the unit of time. This per-unit time change, denoted ΔS, is the rate at which S changes. At any time, molecules are being taken from S and bound with enzyme molecules. Occasionally, the reverse occurs: substrate molecules bound with enzyme molecules are released at random without the reaction taking place. Thus, substrate molecules are lost, and substrate molecules are gained. Any rate like ΔS can be written as a rate of gains minus a rate of losses:

$$S = \{\text{rate of gains}\} - \{\text{rate of losses}\}.$$

Think of putting 3 marbles per second into a large bowl of marbles and taking out 5 per second; the net rate of change (marbles) would be –2 marbles per second.

In a well-stirred chemical soup, the number of collisions between two types of molecules is approximately proportional to the product of their concentrations. We can model the rate of losses of substrate molecules as $c_1 S(E - B)$, that is, as the rate at which substrate and free enzyme molecules collide. If a constant proportion of bound enzyme molecules fail to complete the catalysis and release their substrate molecules in a unit of time, then we can model the rate of gains of substrate molecules as $c_1 B$. Here, c_1 and c_2 are reaction rate constants. Thus, we have established the following equation for the rate at which substrate changes:

$$S = c_2 B - c_1 S(E - B).$$

Next, we construct an equation for the rate ΔB at which the concentration of bound enzyme molecules changes. Any losses to S are gains to B and vice versa, and so the terms in the ΔS equation will enter the ΔB equation, except with the signs reversed. Also, bound enzyme molecules are lost when they complete catalysis and release product molecules. Let us model that loss as a proportional rate $c_3 B$, where c_3 is another rate constant. Thus, the equation for the rate ΔB becomes

$$B = c_1 S(E - B) - c_2 B - c_3 B.$$

A key step in the derivation occurs here. For any given value of S, the value of B should quickly tend to reach an **equilibrium**, a steady state in which gains to B exactly balance losses. If B rises above the equilibrium (more enzyme molecules busy), then fewer enzyme molecules will be available to grab substrates, and the busy molecules will create products at a faster rate. Thus, B will tend to decrease. If, however, B is below the equilibrium, then the fewer busy molecules will form products, and more available enzyme molecules will bind with substrate, causing B to increase. Biochemists call this the "quasi-steady-state hypothesis." If B is in chemical equilibrium, it means that B is not changing. In other words, the rate of change of B is 0:

$$B = c_1 S(E - B) - c_2 B - c_3 B = 0.$$

The equilibrium value of B, denoted as $\bar{B}$, is found by algebraically solving $c_1 S(E - B) - c_2 B - c_3 B = 0$ for B (try filling in the steps):

$$\bar{B} = \frac{c_1 ES}{c_1 S + c_2 + c_3}.$$

The value of the equilibrium for B is seen to depend on the level of S. Now, we can take this expression for $\bar{B}$ and substitute it for B in the equation for S:

$$S = c_2 \bar{B} - c_1 S(E - \bar{B}) = c_2 \frac{c_1 ES}{c_1 S + c_2 + c_3} - \left(E - \frac{c_1 ES}{c_1 S + c_2 + c_3}\right)$$

This is a form of the answer we were pursuing, that is, we wanted an expression for how the rate of change of substrate concentration depends on substrate concentration. With a little patience, we can simplify the expression considerably. Note the term in parentheses. We find that

$$E - \frac{c_1 ES}{c_1 S + c_2 + c_3} = E\left(1 - \frac{c_1 S}{c_1 S + c_2 + c_3}\right)$$

$$= E\left(\frac{c_1 S + c_2 + c_3}{c_1 S + c_2 + c_3} - \frac{c_1 S}{c_1 S + c_2 + c_3}\right) = E\frac{c_2 + c_3}{c_1 S + c_2 + c_3}$$

Put our new form in place of the parenthesized term to find that

$$S = c_2 \frac{c_1 ES}{c_1 S + c_2 + c_3} - c_1 \frac{(c_2 + c_3)ES}{c_1 S + c_2 + c_3} = -\frac{c_1 c_3 ES}{c_1 S + c_2 + c_3}.$$

We are just 10 feet from the water hole. Divide numerator and denominator by c_1 to get

$$S = -\frac{c_3 ES}{S + \left(\dfrac{c_2 + c_3}{c_1}\right)}.$$

This expression for the net rate of change of S is negative and signifies that the processing of substrate into product is consuming substrate. If we are measuring the *amount* of loss per unit time (rate of consumption of substrate), we are measuring $-S$:

$$-S = \frac{c_3 ES}{S + \left(\dfrac{c_2 + c_3}{c_1}\right)}.$$

The equation is now in the familiar form

$$y = \frac{b_1 x}{b_2 + x},$$

where $b_1 = c_3 E$, $b_2 = (c_2 + c_3)/c_1$, x is substrate concentration, and y is the rate of substrate consumption.

The derivation can be repeated for predation by analogy. Think of E as the total gut volume of a wolf, and B as the volume of the gut that is filled with moose meat. "Hunger" is represented by $E - B$, the empty gut volume. The density of moose is S. We assume moose are attacked at a rate $c_1 S(E - B)$ (proportional to moose density $\times$ hunger) and that some attacks fail at a rate $c_2 B$. The gut contents are cleared and made into the product (wolf tissue and metabolism) at a rate $c_3 B$. We have the basic structure for the equations for S and B.

Why should you learn algebra? So that you can understand the world around you! And why should you learn R? So that you can put your understanding to work!

WHAT WE LEARNED

1. Linear statistical models with more than one predictor variable have the form

$$y_{\text{predicted}} = b_1 + b_2 u + b_3 v + b_4 w + \ldots,$$

where u, v, w, $\ldots$ are predictor variables and b_1, b_2, b_3, $\ldots$ are constants with values to be determined by least squares. Each observation in the data for such a model would have a value y_i of the response variable and values u_i, v_i, w_i, $\ldots$ of the predictor variables ($i = 1, 2, \ldots, n$). The number n of observations should exceed the number of predictor variables by at least a factor of 2, preferably more, for obtaining a model with adequate prediction properties. However, the least squares calculations can be performed with the number of predictor variables (counting the column of 1s in the X matrix) less than or equal to the number of observations. The matrices formed with the data are given by

$$X = \begin{bmatrix} 1 & u_1 & v_1 & w_1 & \cdots \\ 1 & u_2 & v_2 & w_2 & \cdots \\ \vdots & \vdots & \vdots & \vdots & \\ 1 & u_n & v_n & w_n & \cdots \end{bmatrix}, \quad y = \begin{bmatrix} y_1 \\ y_2 \\ \vdots \\ y_n \end{bmatrix}$$

The column vector of unknown constants is

$$\mathbf{b} = \begin{bmatrix} b_1 \\ b_2 \\ b_3 \\ \vdots \end{bmatrix}.$$

The least squares estimates of the unknown constants are solutions to a system of linear equations given by

$$(\mathbf{X'X})\mathbf{b} = \mathbf{X'y},$$

which is a matrix expression of a result due to Legendre. The solution can be obtained as

$$\hat{\mathbf{b}} = \begin{bmatrix} \hat{b}_1 \\ \hat{b}_2 \\ \hat{b}_3 \\ \vdots \end{bmatrix} = (\mathbf{X'X})^{-1}\mathbf{X'y}.$$

The `solve()` function in R can perform the calculations. The predictions for every observation are calculated as

$$\mathbf{y}_{\text{predicted}} = \mathbf{X}\hat{\mathbf{b}}.$$

2. A nonlinear statistical model is a prediction equation in which one or more of the unknown constants enters nonlinearly. An example is the kill rate equation used in Chapter 1 to model how the number of moose killed in one unit of time by an average wolf varies with the density of moose:

$$y = \frac{b_1 x}{b_2 + x}.$$

A special function built into R called `nls()` can perform the iterative calculations necessary to fit nonlinear statistical models to data using the criterion of least squares.

Computational Challenges

15.1. A polynomial prediction equation has the form

$$y_{\text{predicted}} = b_1 + b_2 x + b_3 x^2 + \ldots + b_{k+1} x^k.$$

Fit a linear equation ($k = 1$), a quadratic equation ($k = 2$), a cubic equation ($k = 3$), a quartic equation ($k = 4$), and a quintic equation ($k = 5$) to the following data:

x	3	5	8	15	19	22	30	34	36
y	17.8	18.6	32.3	34.7	37.4	33.5	27.8	30.4	29.0

For each model, draw a scatterplot of the data with the fitted model overlaid. What happens to the prediction when there are too many terms in the prediction equation?

15.2. Currently popular among scientists is a "model selection index" called AIC (for Akaike Information Criterion, pronounced ah-kah-EE-kay; see Burnham and Anderson 2002). For any model fitted with least squares, the AIC can be calculated as

$$\text{AIC} = n\left[\log\left(\frac{2\pi\,\text{SS}}{n}\right) + 1\right] + 2(p+1),$$

where SS is the minimized sum of squared prediction errors for the data, n is the number of observations, π is the circle ratio constant ($= 3.14159\ldots$), and p is the number of constants estimated in the model. The procedure for using AIC is to fit all models under consideration and pick the model having the lowest value of AIC. The AIC can be positive or negative. The $2(p+1)$ term is a "penalty" for unknown constants, which is derived to optimize prediction quality among the collection of models being considered.

Using the pronghorn data from Computational Challenge 13.1, fit models to predict fawn count using all the possible subsets of the predictor variables. Calculate the AIC for each model. Which set of predictor variables provides the best predictions, according to AIC?

15.3. Fit the kill rate equation to each of the three sets of lady beetle feeding rate data from Computational Challenge 2.8.

15.4. Below are two sets of data on the reaction rate of an enzyme under two different experimental conditions (treated with Puromycin and untreated). The data were collected by Treloar (1974) and form the preloaded data frame Puromycin in R. Fit the enzyme rate equation separately to the treated and the untreated data. Draw both data sets on the same scatterplot, using different point symbols to distinguish the two treatments. Add plots of the fitted enzyme rate equations to the scatterplot.

conc	rate	state
0.02	76	treated
0.02	47	treated
0.06	97	treated
0.06	107	treated
0.11	123	treated
0.11	139	treated
0.22	159	treated
0.22	152	treated
0.56	191	treated
0.56	201	treated
1.10	207	treated
1.10	200	treated
0.02	67	untreated
0.02	51	untreated
0.06	84	untreated
0.06	86	untreated
0.11	98	untreated
0.11	115	untreated
0.22	131	untreated
0.22	124	untreated
0.56	144	untreated
0.56	158	untreated

15.5. In Computational Challenge 10.6, we saw that Newton's law of cooling is given by the following equation:

$$T = a + (T_0 - a)e^{-kt},$$

where T is temperature of an object at time t, T_0 is the initial temperature of the object, a is the ambient temperature, and k is a cooling rate constant with a value that depends on the properties of the object.

More than 200 years ago, Count Rumford (Benjamin Thompson) performed an experiment to record the cooling of a cannon barrel (Benjamin Count of Rumford 1798). The data are preloaded in R as the data frame Rumford. He heated the barrel to 130°F by grinding it internally with a blunt bore, turned by a team of horses. The ambient temperature was 60°. Rumford monitored the temperature of the cannon with a thermometer for more than 40 minutes. Fit Newton's law of cooling to his data, listed below. The single unknown constant to be determined is k. You will need to devote some thought to how to pick an initial value of k

time (min)	temperature (°F)
4	126
5	125
7	123
12	120
14	119
16	118
20	116
24	115
28	114
31	113
34	112
37.5	111
41	110

for the iterative calculations. Draw a graph of the data and fitted model. In light of your results, comment on the quality and appropriateness of Newton's model for this particular application.

15.6. Computational Challenge 10.9 listed data on Norway's North Sea oil production. Fit the Hubbert's peak curve from Chapter 10 to the oil production data with nonlinear least squares, using the timescale given in the challenge. The unknown constants to be determined are m, k, and b. Computational Challenge 10.9 already provided the nonlinear least squares estimates of m, k, and b, and so the present challenge amounts to verifying the estimates. You can use nearby rounded values as initial values.

Afternotes

If you want to read more about the uses of the diminishing returns model in biology, the equation is called the **Michaelis–Menten model** when used to describe an enzyme reaction rate and the **Holling type II functional response** when used to describe a predation rate. In microbiology, the equation is used to describe the rate at which bacteria take up nutrients (as a function of the nutrient concentration) and is called the **Monod model**. The three applications arose and existed for years almost independently in different subspecialties of biology. Nowadays the phenomena the equation describes are understood to have similar underlying system structures. Good things happen when scientists in different disciplines become more quantitative (and when they talk to each other more). Mathematically, the equation is called a rectangular hyperbola.

References

Benjamin Count of Rumford, F. R. S. M. R. I. A. 1798. An inquiry concerning the source of the heat which is excited by friction. *Philosophical Transactions of the Royal Society of London* 88:80–102.

Burnham, K. P., and D. R. Anderson. 2002. *Model Selection and Multimodel Inference: A Practical Information-Theoretic Approach*, 2nd edition. New York: Springer-Verlag.

Gotelli, N. J. 2008. *A Primer of Ecology*. Sunderland: Sinauer.

Messier, F. 1994. Ungulate population models with predation: A case study with the North American moose. *Ecology* 75:478–488.

Treloar, M. A. 1974. *Effects of Puromycin on Galactosyltransferase in Golgi Membranes*. M.Sc. Thesis, University of Toronto.

16

Conclusion—It Doesn't Take a Rocket Scientist

In this concluding chapter, we are going to put it all together and try something big.

It will be as big as the Earth, as big as the sun, as big as the Earth's orbit around the sun.

Real Solar System Example

We saw in Chapter 9 that Earth's path around the sun has the shape of an ellipse and that the polar curve equation for the ellipse is simple enough to compute and plot, even on a handheld graphing calculator.

This chapter tackles the orbit question differently. Here, we start out with Newton's universal law of gravitation and find out what shape it implies for the Earth's orbit. We will do this by numerically "solving" the gravity equations to obtain directly the projected path of the Earth around the sun. The R script used for this purpose will have to perform many thousands of computations. The script gives a small preview of the types of calculations undertaken by space scientists to predict planetary motions or send an orbiter to another planet.

In this chapter, a moderately large (by the standards of this book) R script for the basic calculations is provided. We will go through all the calculation details. Your task in this chapter is to get the script to run on your computer, and in the computational challenges you will have the opportunity to alter the script in interesting and informative ways.

This chapter uses R like it was meant to be used. Do try this at home. Do not try this on your graphing calculator.

The Problem

Newton's universal law of gravitation states that there is a force of attraction between any two objects having mass. The force decreases with the square of the distance between the two masses and is proportional to the product of the two masses. Newton was able to show mathematically that this law *implies* Kepler's laws of planetary motion. Further, Newton's gravitational law predicts phenomena beyond Kepler's ellipses. For instance, according to Newton's gravity an object with sufficient velocity will not orbit another much larger object but will rather execute a flyby in which its trajectory is bent by the large object's gravity. The flyby solution was fortunate for the astronauts of *Apollo 13*, whose spacecraft was disabled on its way to the moon in 1970. Their spacecraft was whipped around by the moon's gravity, and they returned to Earth safely. Normally, the demonstration of "conic section" solutions (represented by the polar curve from Chapter 9) to Newton's gravity interactions for two bodies—ellipses, parabolas, and hyperbolas—requires an engineering-level physics course in college.

With R, however, we do not have to wait. We can simply calculate. We can divide the Earth's trajectory into thousands of tiny changes in the Earth's position and accumulate the resulting positions in vectors. However, to calculate anything we must take the time to set up the problem precisely.

Let us first set up a coordinate system for the Earth and the sun (Figure 16.1). We designate the sun to be the origin, and let us suppose that the coordinate (x, y) is the current location of Earth. We denote the mass of sun by M and the mass of Earth by m (we will fill in the numerical values for these constants later, in Outline of R Script for Calculating the Trajectory of Earth.).

Our goal is to calculate the trajectory of the Earth around the sun using just Newton's gravitational law. We will not need any fancy mathematics; with R, we can do the calculation by brute force. The strategy is to use Newton's gravity to calculate how the Earth's position (values of x and y) changes over a tiny time interval (a few hours). We will then update the Earth's position with the new values of x and y, and we will turn around and calculate another position change for another tiny time interval. We merely need to repeat this process for enough tiny time intervals to accumulate to, say, a year into the future ("we?" your computer mutters under its breath, somewhat disgustedly).

The main challenge is the bookkeeping involved. The calculation will be long and detailed, and we must be thoroughly organized. In our R script,

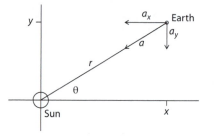

FIGURE 16.1
Coordinate system for the Earth's position relative to the sun.

some things have to be calculated first and some things must be calculated later. Before starting to write computer code, we must lay out all the tasks carefully and in the right order.

The Concept

First, we must understand the calculation. In order to calculate the extent to which x and y will change, we need to know the velocities in the x and y directions. Let us denote the velocities v_x and v_y, respectively. If t is time and t a tiny interval of time, we calculate a change in x by the product v_x t and a change in y by the product v_y t. However, not only will the values of x and y change during every tiny time interval, but the velocities of Earth in those directions will also change. Gravity is a force, which means it changes the velocity of an object. The force will have a different value for every different distance of Earth from the sun. For each tiny time interval, we will have to calculate how much v_x and v_y change, in addition to the changes in x and y.

Changes in Velocities

Gravity acts as a force of attraction between two masses. The sun is much more massive than Earth and is hardly budged by the Earth's pull. The force exerted by the sun on the Earth, however, is substantial in comparison to Earth's mass. According to Newton's second law of motion, any force that would change Earth's motion can be written as follows:

$$F = ma,$$

where F is the force and a is the acceleration (the rate at which Earth's velocity changes). But also, according to Newton's universal law of gravitation, the gravitational attraction force between the sun and the Earth, denoted by F_{Mm}, obeys the famous inverse-square relationship:

$$F_{Mm} = -G\frac{Mm}{r^2}.$$

Here, r is the distance between the (centers of the) sun and the Earth, and G is the universal gravitation constant. For gravity, these laws are expressions of the same force, that is,

$$F = F_{Mm},$$

or

$$ma = -G\frac{Mm}{r^2}.$$

Cancelling the m's, we see that acceleration of the Earth due to the sun's gravitational attraction is

$$a = -G\frac{M}{r^2}.$$

This acceleration is always in the direction of the sun. The above expressions actually represent a simplified version of Newton's laws. The above expressions are written for a one-dimensional coordinate system. The minus sign designates that the gravity acceleration tends to decrease the position of the Earth on a line connecting the Earth and the sun. The modern-day, general formulation of Newton's laws uses concepts from vector calculus in order for the formulation to be independent of any coordinate system. To know about this general formulation, you must wait for your college engineering physics course.

For now, we are interested in calculating the Earth's path in our coordinate system shown in Figure 16.1. To apply the gravitation law in our coordinate system, we must translate the force toward the sun into component forces in our x and y directions.

In our coordinate system, gravity acceleration has a component in the x direction and a component in the y direction. In Figure 16.1, the arrow labeled a depicts the gravitational acceleration pulling the Earth in the direction of the sun. From Figure 16.1 and the definitions of the trigonometric functions discussed in Chapter 9, we can understand that the x and y direction components of Earth's acceleration are given by

$$a_x = a\cos\theta,$$

$$a_y = a\sin\theta,$$

where θ is the angle made by a line connecting the sun and the Earth with the line given by the x-axis of our coordinate system.

But we can express a, $\cos\theta$, $\sin\theta$, and r in the above components all in terms of the position coordinates x and y. Noting that

$$r^2 = x^2 + y^2 \quad \text{(Pythagoras theorem)},$$

$$r = \sqrt{x^2 + y^2} = \left(x^2 + y^2\right)^{1/2},$$

$$\cos\theta = \frac{x}{r} \quad \text{(definition of cosine)},$$

$$\sin \theta = \frac{y}{r} \quad \text{(definition of sine)},$$

we can see that

$$a_x = a \cos \theta = \left(-\frac{GM}{r^2}\right)\left(\frac{x}{r}\right) = -\frac{GMx}{\left(x^2 + y^2\right)^{3/2}},$$

$$a_y = a \sin \theta = \left(-\frac{GM}{r^2}\right)\left(\frac{y}{r}\right) = -\frac{GMy}{\left(x^2 + y^2\right)^{3/2}}.$$

It is these current accelerations in the x and y directions, calculated for the current location of the Earth (x and y), that allow us to calculate changes in the velocities v_x and v_y. The amount of change in v_x during t is the product of the rate at which v_x changes and t; but the rate at which v_x changes is just the acceleration component a_x:

$$v_x = a_x \ t.$$

Here v_x is the amount of change in v_x. Similarly, for the y direction we have the change in v_y, given by

$$v_y = a_y \ t.$$

Move the Earth

Our core task, remember, is to calculate the change in the position of Earth over a small interval of time. All the pieces are in place. Let us summarize the necessary calculations. We start with the current numerical values of x, y, v_x, v_y, and t. Calculate the changes in position coordinates as follows:

$$x = v_x \ t,$$

$$y = v_y \ t.$$

Calculate the changes in velocity components as follows:

$$v_x = -\frac{GMx}{\left(x^2 + y^2\right)^{3/2}} \ t,$$

$$v_y = -\frac{GMy}{\left(x^2 + y^2\right)^{3/2}} \ t.$$

We now can move the Earth to its new location. After we have computed the above changes to x, y, v_x, and v_y, the new values of x, y, v_x, v_y, and t are calculated as the old values plus the changes:

$$x_{NEW} = x + \quad x,$$

$$y_{NEW} = y + \quad y,$$

$$v_{x\ NEW} = v_x + \quad v_x,$$

$$v_{y\ NEW} = v_y + \quad v_y,$$

$$t_{NEW} = t + \quad t.$$

We then start the calculations anew using the new values of coordinates, velocities, and time.

Getting Organized

It should strike you that repeatedly calculating changes in the Earth's location is wonderfully suited to a for-loop in R. Indeed, such a loop will be the heart of the R script. But before we can calculate anything, the script must be given the numerical values of G and M; the starting values of x, y, v_x, and v_y; and the value of t. We must also make decisions about what measurement units to use for distance, time, and mass. Moreover, we must creatively hit on the key idea for writing the R script: We will fill two large vectors, say, x and y, with the successive values of positions x and y calculated in the loop. Then we just ask R to do a line plot of x and y.

For measurement units, a nice distance scale for the plot is AU. Note that 1 AU is the average distance of the Earth from the sun (about 149,597,871 km or 92,955,807 mi). As the unit of time, a sidereal year is handy. A sidereal year is the time it takes Earth to make one complete revolution around the sun, from periapsis to periapsis (periapsis, also known as perihelion, is the point at which the Earth's orbit is closest to the sun). The sidereal year is roughly 20 minutes longer than our more familiar tropical year (solstice to solstice, on which our calendar is based) due to the slow precessional wobbling of the Earth's axis of rotation. For mass, we can conveniently use the mass of the Earth as 1 unit.

Various physical and astronomical quantities that we need, such as G and the velocity of Earth at periapsis, are listed in books or given online in a variety of measurement units. We must convert the units of all such quantities to the units we have settled on for the script. It seems convenient to do the conversion of units in the script itself.

What follows is, first, an outline and discussion of the tasks the script has to perform (Outline of R Script for Calculating the Trajectory of Earth section) and, second, the script itself (The R Script section). The numbered headings in the outline correspond to numbered sections in the script. Study each section of the script, line by line. Make sure you understand what every line of the script does.

Outline of R Script for Calculating the Trajectory of Earth

0. Set values for the initial location, velocity, and direction of Earth.

We can take advantage of the fact that we already know much about the Earth's orbit to set these quantities to values that result in a nice plot. We take the major (long) axis of the elliptical orbit to lie on the *x*-axis (recall that the sun is the origin). The periapsis is at a point slightly less than 1.0 on the *x*-axis; the value of *y* here is of course 0. The periapsis point will be the initial position of the Earth. The Earth will be crossing the *x*-axis vertically at the initial instant in the direction increasing for *y*. In the script, we put in an initial angle of crossing, called ϕ (phi), with a value set at $\pi/2$ (vertical, in radians). We thought it might be fun to mess around with this angle later, to see what would happen if Earth was "thrown" in a different direction. The initial velocity is taken as the velocity of Earth at periapsis, as cataloged in online astronomy tables.

1. Set the total duration of time for which the trajectory is to be calculated and the total number of tiny time intervals per sidereal year to use.

We will start with the total time duration of 1 sidereal year. If we start exploring slower orbits of outer planets later, we will want to set the total time duration to be longer. The number of tiny time intervals per sidereal year (named n.int in the script) is taken as 10,000, which should be enough for a good graph of the shape of the orbit. A larger number of tiny time intervals per sidereal year should result in a more accurate trajectory, but then this would also make us wait a few seconds longer for the script to run.

2. Set values for the physical constants needed in the calculations.

Thank you, Wiki P.

3. Rescale the physical constants to consistent and computationally easy units.

For instance, the usual units of *G* are meters³ kilograms⁻¹ seconds⁻². Our unit of distance is 1 AU, our unit of mass is 1 earth mass, and our unit of time is 1 sidereal year. We have to convert *G* to new units for our calculations using the following expression:

$G_{NEW} = G$ (meters per AU)⁻³(kilograms per earth mass)(seconds per sidereal year)²

This will be on the test.

4. Calculate the initial quantities needed.

Calculate the initial velocities in the *x* and *y* directions, size of the tiny time interval to be used, and total number of time intervals (named tot.int in the script) to be used. The quantity `tot.int` is different from the number of tiny intervals (`n.int`) in 1 sidereal year. We can change `tot.int` when we want to end the trajectory at some time other than 1 sidereal year.

5. Set up vectors to store the x and y coordinates, the velocities v_x and v_y, and time t.

We "preallocate" the numeric vectors that will store all our results. The vectors will then exist in the computer's memory, waiting to be filled. The vectors will start out with all elements equal to zero, but the script will subsequently change those values. Each vector has a length equal to the total number of tiny time intervals + 1. The extra element is needed to fill the initial value.

6. Initialize the first value in each vector.

Here, the user-provided values from part 0 in the script are inserted into the vectors in preparation for the big loop calculation. Also, the product GM occurs in the velocity change equations, and it is faster to calculate the product just once outside the loop rather than over and over inside the loop.

7. Loop to calculate trajectories.

The for-loop will execute the R commands between the { and } symbols over and over again. During the first time through the loop, i will have the value 1; during the second time through, i will have the value 2; and so on until the last time through the loop when i will have the value `tot.int`. Each time through the loop, the equations will be accessing different elements of the vectors, using i in the index numbers of the vectors. There are 9 commands inside the loop:

7a: Calculate the change in x.

7b: Calculate the change in y.

7c: Calculate the change in v_x.

7d: Calculate the change in v_y.

7e: Calculate and store the new value of x.

7f: Calculate and store the new value of y.

7g: Calculate and store the new value of v_x.

7h: Calculate and store the new value of v_y.

7i: Calculate and store the new value of t.

During the last time through the loop when i has the value `tot.int`, the script will be calculating the value of element `tot.int+1` in each of the vectors. In particular, the position vectors x and y will be completely filled with the successive calculated positions of Earth.

8. Plot y positions versus x positions in a simple line plot, similar to the polar curves given in Chapter 9.

Are you ready for the script?

The R Script

```
#================================================================
#  R program to calculate and plot Earth's orbit around the
#  sun, using Newtonian gravity. Sun is at the origin,
#  x and y coordinates are measured in astronomical units.
#  Time is measured in sidereal years.
#================================================================


#----------------------------------------------------------------
# 0. Set initial location, velocity and direction of the Earth
#    here.
#----------------------------------------------------------------
x0=.983291      # Periapsis distance in astronomical units.
y0=0            # Start on x-axis, at periapsis.
phi=pi/2        # Initial direction angle of earth's orbit.
v0=30.3593      # Earth's initial velocity at periapsis, km/s.


#----------------------------------------------------------------
# 1. Set number of time increments and end time for trajectory
#    calculation.
#----------------------------------------------------------------
n.int=10000     # Number of time increments per Earth year
                #   to be used.
t.end=1         # Number of years duration of trajectory.


#----------------------------------------------------------------
# 2. Various constants gleaned from astronomy books or online.
#----------------------------------------------------------------
G=6.67428e-11       # Gravitational constant, m^3/(kg*s^2).
m=6.0e24            # Mass of the earth, kg.
M=m*3.33e5          # Mass of the sun.
km.per.AU=149.598e6 # One astronomical unit, km.
sec.per.year=31558149.540 # Seconds in a sidereal year.


#----------------------------------------------------------------
# 3. Do all calculations in units convenient for plotting.
#----------------------------------------------------------------
me=m/m          # Mass of earth expressed in earth units.
Me=M/m          # Sun's mass in earth units.
```

```
G=G*m*sec.per.year^2/((km.per.AU^3)*(1000^3))
                                    # New units of G are AU^2/
                                    #   (me*yr^2).
v0=v0*sec.per.year/km.per.AU        # New units of velocity are
                                    #   AU/yr.



#------------------------------------------------------------------
# 4. Initialize various quantities.
#------------------------------------------------------------------
v.x0=v0*cos(phi)        # Earth's velocity, x-component, at
                        # periapsis.
v.y0=v0*sin(phi)        # Earth's velocity, y-component, at
                        # periapsis.
dt=1/n.int              # Duration of one tiny time interval
                        # (delta t).
tot.int=round(t.end*n.int);  # Total number of tiny intervals
                        #   (must be an integer).



#------------------------------------------------------------------
# 5.  Pre-allocate vectors which will store all the values.
#     Note: x is x-position, y is y-position, v.x is
#     x-direction velocity, v.y is y-direction velocity,
#     t is time.
#------------------------------------------------------------------
x=numeric(tot.int+1)    # x starts as a vector of ninc+1
                        #   zeros. The "plus one" is
                        #   for the initial condition.
y=x             # Allocate y the same as x.
v.x=x           # Allocate v.x the same as x.
v.y=x           # Allocate v.y the same as x.
t=x             # Allocate t the same as x;



#------------------------------------------------------------------
# 6.  Insert the initial conditions into the vectors.
#------------------------------------------------------------------
x[1]=x0         # Initial value of x-position.
y[1]=y0         # Initial value of y-position.
v.x[1]=v.x0     # Initial value of x-direction velocity.
v.y[1]=v.y0     # Initial value of y-direction velocity.
t[1]=0          # Initial value of time.
c=G*Me          # Pre-calculate a constant that appears
                #   repeatedly in the equations.



#------------------------------------------------------------------
# 7. Loop to calculate trajectory.
#------------------------------------------------------------------
```

```
for (i in 1:tot.int) {
    dx=v.x[i]*dt        # Change in x.
    dy=v.y[i]*dt        # Change in y.
    dv.x=-c*x[i]/(x[i]^2+y[i]^2)^(3/2)*dt    # Change in
                                             #    x-velocity.
    dv.y=-c*y[i]/(x[i]^2+y[i]^2)^(3/2)*dt    # Change in
                                             #    y-velocity.
    x[i+1]=x[i]+dx           # New value of x.
    y[i+1]=y[i]+dy           # New value of y.
    v.x[i+1]=v.x[i]+dv.x    # New value of x-velocity.
    v.y[i+1]=v.y[i]+dv.y    # New value of y-velocity.
    t[i+1]=t[i]+dt          # New value of time.
}


#-----------------------------------------------------------------
# 8. Plot the trajectory.
#-----------------------------------------------------------------
par(pin=c(4,4))     # Equal screen size in x and y directions.
plot(x,y,type="l",xlim=c(-1.1,1.1),ylim=c(-1.1,1.1))
# Line plot of y vs x.
```

Computational Challenges

16.1. Enter the script in R on your computer, and run it. If you are typing, you do not have to type all the comments (the script will run without any of them), but the comments will be helpful if you want to save and use the script in the future. Describe the visible shape of the trajectory in the resulting graph. If desired, alter the script to add the graphical enhancements suggested in Computational Challenge 13.7, and thereby reproduce the figure depicted on the cover of this book.

Once the script runs successfully, save a master copy of it and a copy of the graph. Make a new copy of the script under a new name for alterations.

16.2. Alter the initial *y* direction velocity vy0 of the Earth in the script, and run the modified script. In this and other script modifications, the axis limits for the graph might need to be changed to depict the trajectory well. What happens if this velocity is a little larger, a little smaller, a lot larger, and a lot smaller?

NOTE: You do not have to think of this calculation as throwing the Earth with a different velocity. Instead, you can think of this calculation as what you would do if you were starting a spacecraft from the specified initial location, at velocity vy0, in the absence of the Earth. You will recall that the mass of Earth does not appear in any of the velocity and position change equations used in the script. The same trajectory calculations apply to the spacecraft if the sun is the only nonnegligible source of gravity. Explore a variety of different initial velocities to see what would happen to the rocket.

16.3. Add an initial velocity component in the x direction by altering the initial angle φ (phi in the script). Our spacecraft is now not just crossing the horizontal axis perpendicularly but crossing at a different angle, toward the sun or away from it.

16.4. Set the initial location to x0=1/sqrt(2) and y0=1/sqrt(2), that is, the distance r is exactly 1 AU. "Drop" the Earth by setting its initial velocity to zero. How long will it take for the Earth to reach the surface of the sun (the sun's radius is approximately 0.004652 AU)?

16.5. Obtain the initial conditions for other planets using some Internet research. Plot the orbits of other planets. Note that the amount of Earth years for other planets to complete their orbits are all different; one will have to change the end time for the calculations to get nice graphs.

16.6. Recall from Chapter 9 that a good way to draw a circle is to make a vector of angles, say theta, with elements varying from 0 to 2π, along with a corresponding vector of fixed radii, say, r. Then the vectors defined by the transformation

```
x=r*cos(theta)
y=r*sin(theta)
```

produce the desired vectors x and y containing the circle for plotting. A perfect circular orbit for the Earth on our graph would have r=1, representing an orbit radius of 1 AU.

Alter the R script by adding expressions to compute two new vectors of coordinates on this circle. Alter the script to superimpose the circular figure on the same graph as the Earth's orbit using a different line type, perhaps a dashed line type. Compare the two orbits.

16.7. The actual equation for a trajectory initiated from a point on the positive x-axis at distance r_0 from the origin with an initial velocity of v_0 in the y direction (vertical), written in terms of the distance r and the angle θ, is

$$r = r_0 \frac{(1+\varepsilon)}{(1+\varepsilon\cos\theta)},$$

where $\varepsilon = \left[r_0 v_0^2 / (GM) \right] - 1$. This equation is the famous conic section equation that is the solution for planetary trajectories governed by Newtonian gravity. We graphed this equation in Chapter 9. Alter the R script to overlay a graph of this exact equation on top of the graph resulting from the brute-force calculation. Compare the two orbits.

16.8. Examine the starting and end points of Earth on the original graph, and from the position vectors obtain the numerical coordinates for the starting and end points. Do they match? Our brute-force calculation is only an approximation to the "true" solution to the equations. Can the approximation be made better? See what happens if the number n.int of tiny time intervals is made larger in the original script. A value of 100,000 (why stop there? Try 1,000,000) is not unreasonable.

16.9. (This is a thought challenge.) In Chapter 9, the equation for a projectile (in that case, a thrown baseball) was given to be a parabola. Shouldn't the projectile equation really be the equation for a portion of an ellipse? Resolve this paradox.

Afternotes

Calculus and Conic Sections

The demonstration that the conic section equation (given in Computational Challenges 16.7) is the solution to Newton's gravity equations for a two-body problem is a substantial calculus problem. The student usually encounters it for the first time in a college-level engineering physics course, which is taken after three semesters of calculus. The usual presentation is in polar coordinates, which makes the formulas much less messy.

Feynman's Lost Lecture

Isaac Newton first presented the conic section arguments in purely geometrical terms. The geometrical style of the mathematical argument used by Newton in his masterwork, *The Principia* (Newton 1687) has fallen out of fashion, and much of the treatise remains difficult to read even for modern-day physicists. The twentieth-century physicist Richard Feynman reconstructed, simplified, and refurbished Newton's geometrical derivation of the elliptical trajectory and presented the results in a freshman physics lecture at Caltech, Pasadena, California, in the early 1960s. Feynman's lecture notes on the topic were fleshed out and published, along with an audio recording of the actual lecture, in a book/compact disc (CD) set titled *Feynman's Lost Lecture: The Motion of Planets Around the Sun* (Goodstein and Goodstein 1996). Feynman's lecture is an elegant and highly readable derivation of the elliptical trajectory that uses no calculus; it uses just high-school geometry.

Three-Body Problem

The trajectory equation for two gravitating bodies is an elegant mathematical solution, a thing of beauty. For three or more bodies, however, a simple equation for the trajectory of any of the bodies has never been found. Instead, the gravitational equations must be numerically solved by brute computational force, using an approach similar to the R script discussed in this chapter.

Simultaneous changes in the positions and velocities of all gravitating bodies must be calculated. Each body experiences a gravity force component in the direction of each other body. The resulting trajectories can deviate from the Newton/Kepler ideals in complex ways.

Neptune

Early in the history of planetary studies, astronomers found that the planets exert enough gravitational influence on each other to be detected by attentive observation. In one of the great triumphs of scientific history, the French mathematician Le Verrier used deviations of the orbit of planet Uranus from its ideal to predict the location of an undiscovered large planet with an orbit farther out than that of Uranus. The existence of the new planet, named Neptune by Le Verrier, was confirmed by astronomers in 1846 and was located within 1° of where Le Verrier predicted it would be. In those days, brute-force numerical calculations of solar system motions were done by hand (without calculators) using various clever approximations.

Propagation of Error

Even though careful observational measurements can allow astronomers to determine the positions and velocities of solar system objects with great accuracy, tiny uncertainties always remain due to the limitations of measurement devices. Such uncertainties can be magnified by gravitational trajectories, especially in interactions among three or more bodies. Such uncertainties must be carefully propagated through the trajectory calculations. One error propagation method is to pick out the initial position and velocity at random from a probability distribution (centered at the measured value) using a computer and calculate the trajectory, repeating the pick/calculate process thousands of times. The result is a "band" of possible locations for the object at some future time.

Apophis

The asteroid Apophis was discovered in 2004 in a position about 17 million miles from Earth. It is on a course that will execute a near-miss of our planet in 2029 and return 7 years later. The asteroid is about 1000 feet wide and weighs at least 50 million tons. A collision with Earth would be catastrophic. For instance, if Apophis landed in the Pacific Ocean, the resulting tsunami would destroy all Pacific coastal cities. Although astronomers are confident that the asteroid will skim by Earth harmlessly in 2029, they cannot rule out a collision in the subsequent visit. Astronomers currently calculate that Apophis has a 1 in 250,000 probability of slamming into our planet on Easter Sunday, April 13, 2036. If Earth is lucky and dodges this rock in 2036, there will be additional risky revisits of Apophis to Earth's vicinity during the remainder of the century, as by that time the asteroid will have become locked into a dangerous orbital dance with our planet. Have a nice day.

Orbit of Pluto Is Chaos

The orbit of Pluto is highly elongated, with a periapsis closer to the sun than the planet Neptune. The gravitational influence of Neptune and the other outer planets is enough to affect Pluto in such a way that the orbit of Pluto never quite repeats itself. The orbit is "chaotic," an ongoing wobbling loop that misses its starting point slightly each time around.

The *m*s Cancel

A pebble or the Earth initiated to move with the same position and velocity would trace the same orbit around the sun, according to Newtonian gravity.

Mercury Orbit Precession, and General Relativity

The planet Mercury's orbit is different. The periapsis point wanders slowly around the sun, giving a precession to the orientation of the orbit. If traced for a long enough time, the orbit would form sort of a neat spirograph pattern. In the 1800s, astronomers realized Mercury's discrepancy from the Newtonian two-body elliptical form, but they were unable to account for the degree of the discrepancy by including influences of other planets as well as the sun. In 1915, Albert Einstein proposed the theory of general relativity, which states that there is no attraction force between masses; rather, a mass bends the space and time in its vicinity. Mercury is falling freely through space and time that is bent near the sun. The theory predicted the precession of Mercury's orbit almost exactly. In 1919, astronomical expeditions to Africa and South America to observe a total eclipse of the sun announced that light from stars positioned behind the sun was bent while traveling past the sun toward Earth. Newton's gravity was overturned, and Einstein became an international celebrity almost overnight. Today, the accuracy of your global positioning system (GPS) unit depends critically on correcting the time signals from GPS satellites with calculations from general relativity, because the space–time of the satellites is bent by Earth.

Measurement Units

The following is a press release from National Aeronautics and Space Administration (NASA) concerning the loss of the $328-million-dollar *Mars Climate Orbiter* spacecraft in 1999.

MARS CLIMATE ORBITER TEAM FINDS LIKELY CAUSE OF LOSS

A failure to recognize and correct an error in a transfer of information between the Mars Climate Orbiter spacecraft team in Colorado and the mission navigation team in California led to the loss of the spacecraft last week, preliminary findings by NASA's Jet Propulsion Laboratory internal peer review indicate.

"People sometimes make errors," said Dr. Edward Weiler, NASA's Associate Administrator for Space Science. "The problem here was not the error, it was the failure of NASA's systems engineering, and the checks and balances in our processes to detect the error. That's why we lost the spacecraft."

The peer review preliminary findings indicate that one team used English units (e.g., inches, feet and pounds) while the other used metric units for a key spacecraft operation. This information was critical to the maneuvers required to place the spacecraft in the proper Mars orbit.

"Our inability to recognize and correct this simple error has had major implications," said Dr. Edward Stone, director of the Jet Propulsion Laboratory. "We have underway a thorough investigation to understand this issue."

NASA PRESS RELEASE 99-113, September 30, 1999 (NASA 1999)

NASA engineers should have known it would be on the test. Uh, maybe it takes a rocket scientist after all. Maybe you!

References

Goodstein, D. L. and J. R. Goodstein. 1996. *Feynman's Lost Lecture: The Motion of Planets Around the Sun*. New York: W. W. Norton & Co.

NASA. 1999. "Mars Climate Orbiter Team Finds Likely Cause of Loss." http://www.nasa.gov/home/hqnews/1999/99-113.txt. Accessed June 28, 2012.

Newton, I. 1687. *Philosophiae Naturalis Principia Mathematica*. Londini: jussi Societatus Regiae ac typis Josephi Streater; prostat apud plures bibliopolas.

Appendix A: Installing R

When you visit http://www.r-project.org/, take some time to browse the list of links on the left side of the web page. Read "About R" and get a feel for the purpose, history, and spirit of R and the R project.

Further, the "FAQs" (frequently asked questions) contain much valuable information about installing R. Things you will want to check are which version of R best matches your computer and operating system, and the installation procedures you will use. Get an overview of what to expect during the installation process. Read the FAQs for your particular operating system, be it Windows, Mac OS, or Unix/Linux.

If you have installed software packages on your computer before, you should have no difficulties with installing R. If your operating system is Windows, you should determine if it is a 32- or 64-bit operating system. In the Start menu, click Computer and then System Properties. If your operating system is 64 bit, then it can run either the 32-bit or the 64-bit versions of R, and I recommend you to install the 64-bit version of R. If you are working in Linux, use of a package management system such as Yellowdog Updater, Modified (Yum) provides the easiest installation.

Two main problems can be encountered: (1) You are not the administrator or do not have installation privileges on the computer. If it is an institutional computer, then consult your information technology administrator for installation. Remind the administrator that R has no license fee. (2) Your computer or operating system, or both, is old. You might have to select an earlier version of R to install. Then you must know the version of your operating system in order to pick the right older version of R to match it during the installation process. The FAQs will advise you about the earlier R versions.

When you are ready, you will click on "CRAN" (comprehensive R archive network). The network is a series of servers all over the world from where the latest version as well as earlier versions of R can be downloaded and installed. From the list of server links, choose a server near you and click on it.

Once connected to an R mirror server, you will see the option "Download and Install R." Pick the link with the appropriate operating system version and click on it. This will begin the process of downloading the precompiled binary version of R for your operating system, which will be more than fine for most things you will ever do in R. Further, for the purposes of this book and for most of your initial purposes, you will need only the base R package (the contributed packages are routines for advanced analyses that have been written by scientists around the world).

In Windows, the downloader will ask you whether you want to run or save the executable (.exe) file; choose "run." The Windows installer will open and the installation process will proceed. You will normally want to select all the default options for installation (and the 64-bit version if your system is 64 bit).

In the Mac OS process, the binary file gets downloaded to your system. It is an installer package, and you have to find it and double-click on it to start the installation rolling.

For Linux, there are R binaries available for Debian, Red Hat, and Ubuntu. For other types of Unix, you might have to compile your R package from the source code. The R FAQs are the place to begin installation in this case.

Appendix B: Getting Help

For help concerning a known R command or function, you can use the `help()` function. For instance, to get a listing of information about the `plot()` function, one can type the following at the console:

```
> help(plot)
```

If you need help with a topic—perhaps you are looking for the name of a function in R that will provide what you are looking for—type a key word after two question marks:

```
> ??histogram
```

For more help, there is the `Help` menu on the console menu bar. Among other things, you will find two PDF manuals: (1) `An Introduction to R`, and (2) the `R Reference Manual`. The R reference manual (The R Development Core Team. 2010. *R: A Language and Environment for Statistical Computing, Reference Index.* R Foundation for Statistical Computing) is the comprehensive catalog of R commands. The sheer amount of preexisting functions at your disposal is spectacular. Browse the sections on "base," "graphics," and "statistics" for ideas on how to tackle your particular problem.

Finally, if you have a question about R, it is more likely than not that the same question has already occurred to someone else in the world. Just type your question into an Internet search engine, for instance, "convert logical vector to numeric vector in R." You will likely find your question answered in many online forums in many creative ways (in this case, for the curious, add 0 to the logical vector or multiply it by 1).

Appendix C: Common R Expressions

Arithmetic Operations

All arithmetic operators work elementwise on vectors. The priority of performing operations is ^, * and /, and + and –, and priority goes from left to right.

x+y: Addition.

x–y: Subtraction.

x*y: Multiplication.

x/y: Division.

x^y: Power. Example: 2^3 is 2^3.

( ): Give priority to operations in parentheses.

Example:

```
> x=3*(4-5^2)+(12/6*2+5*(4-2))
> x
[1] -49
```

Vector Operations

c(a,b,c, ...): Combine the scalars or vectors a, b, c, ... into a vector.

numeric(n): Vector of 0s of length n.

length(x): Number of elements in vector x.

n:m: Index vector n, n+1, n+2, ..., m.

n:-m: Index vector n, n–1, n–2, ..., –m.

seq(n,m,k): Additive sequence n, n+k, n+2k, ..., m.

rep(x,n): Returns a vector with x repeated n times.

x%*%y: Dot product of vectors x and y (each with n elements): $x_1y_1 + x_2y_2 + \cdots + x_ny_n$.

Assignment Statements

= (or <-): Evaluate what is on the right and store it under the name on the left.

Examples:

```
> x=c(3,4,5)
> x
[1] 3 4 5
> y=x-2
> y
[1] 1 2 3
> z<-x*y
> z
[1]  3  8 15
```

Logical Comparison Operators

x>y, x<y, x>=y, x<=y, x==y, x!=y: Greater than, less than, greater than or equal to, less than or equal to, equal to, and not equal to. Compares elements of x and y and returns the logical vector (TRUE TRUE FALSE ...).

&: And operator. Example: (x>=y)&(x<=z).

|: Or operator. Example: (x>=y)|(x<=z).

!: Not operator. Example: !(x>=y).

Element Selection and Manipulation

Selecting with Index Numbers

x[j]: The jth element of x

x[-j]: All elements of x except the jth element

x[c(1,3,4)]: First, third, and fourth elements of x

x[n:m]: The nth through mth elements of x

Selecting with Logical Vectors

x[TRUE FALSE TRUE TRUE FALSE ...]: All elements in x corresponding to TRUE.

Selecting with Logical Comparison Operators

x[a<b]: Elements of x corresponding to element in a and b for which comparison is true.

x[x<7]: Elements of x that are strictly less than 7.

x[(x>=2)&(x<=8)]: Elements of x that are between 2 and 8 including 2 and 8.

which.max(x): Returns index of the greatest element of x.

which.min(x): Returns index of the smallest element of x.

which(x==a)   (which(x<=a), etc.): Returns vector of indices of x for which comparison operator is true.

rev(x): Reverse the elements of x from last (x[n]) to first (x[1]), where n=length(x).

sort(x): Sort the elements of x from smallest element to largest element.

replace(x,k,y): Replace elements of x designated by index vector k with corresponding elements of y.

na.omit(x): Omit missing values (data code: NA) from vector or lines with missing data from data frame.

Scripts

Scripts are plain text files of R commands, traditionally named with the extension ".R." Current versions of R for Windows and Mac operating systems have a built-in R editor; for current Unix/Linux versions, a separate text editor is needed.

In the operating systems Windows or Mac OS, scripts are run from a pull-down menu, or by highlighting portions and using the pull-down menu, or by copy/pasting a script into the R console, or by using the source() command.

In Unix/Linux, scripts are run by copy/pasting scripts into the R console or using the source() command.

setwd("c:/R _ stuff/Math"): Set working directory for R to the indicated directory (working directory can also be set from the pull-down menu in the console).

source("coolscript.R"): Run the script named coolscript.R found in R's working directory.

source("c:/R _ stuff/Math/coolscript.R"): Run the script in the indicated directory.

Mathematical Functions

sqrt(x): Square roots of the elements of x.

exp(x): Value of e^x for every element of x.

sign(x): Returns −1, 0, or 1 for negative, zero, or positive element, respectively, of x.

log(x), log10(x), log2(x), log(x,b): Natural (base e) logarithm, base 10 logarithm, base 2 logarithm, and base b logarithm of x.

sin(x), cos(x), tan(x), sec(x), csc(x), cot(x): Sine, cosine, tangent, secant, cosecant, and cotangent of x.

asin(x), acos(x), atan(x), atan2(x,y): Arcsin (inverse sine), arccos, arctan, two-argument arctan2 of x.

factorial(x): Factorial function, that is, $x! = x(x-1)(x-2)\cdots(1)$ with $0! = 1$.

lfactorial(x): $\log(x!)$

gamma(x): Gamma function $\Gamma(x)$ (continuous version of factorial: $\Gamma(x) = (x-1)!$ if $x = 1, 2, 3, \ldots$).

lgamma(x): Logarithm of gamma function: $\log \Gamma(x)$.

max(x), min(x): Largest element, smallest element of x.

range(x): Range of elements of x: max(x)−min(x).

sum(x): Sum of elements of x.

prod(x): Product of elements of x.

cumsum(x): Cumulative sum of elements of x.

cumprod(x): Cumulative product of elements of x.

pmin(x,y,z, ...), pmax(x,y,z, ...): Pick minimum and pick maximum from vectors x, y, z, and return as vector.

diff(x): First differences of elements of x: x[2:n]−x[1:(n−1)] where n=length(x).

mean(x): Sample mean of elements of x: sum(x)/length(x).

var(x): Sample variance of elements of x: sum((x-mean(x))^2)/(length(x)−1).

sd(x): Sample standard deviation of elements of x: sqrt(var(x)).

median(x): Sample median of elements of x.

rank(x): Ranks of elements of x (1 for lowest, averaged ranks for ties).

round(x,n): Round the elements of x to n decimals (n omitted rounds to nearest integer).

floor(x): Round the elements of x down to nearest integer.

ceiling(x): Round the elements of x up to nearest integer.

User-Defined Functions

```
myfunction=function(arg1,arg2,...,argk) {
   statement 1
   statement 2
        :
   return(object)
}
```

myfunction: Name of function arg1, arg2, ..., argk: arguments of function; statement 1, statement 2, ...: list of R statements; object: values or object(s) returned by the function

Example:

```
qdratic=function(a,b,c,x) {
   y=a*x^2+b*x+c
   return(y)
}
A=-1
B=1
C=1
X=seq(-1,2,.1)
qdratic(A,B,C,X)
```

Data Input and Output

The R functions for data input and output have many optional arguments. See the R reference manual (described in Appendix B) for the full listing of possibilities.

setwd("c:/R _ stuff/Math"): Set working directory for R to the indicated directory (working directory can also be set from the pull-down menu in the console).

Mydata=read.table("data.txt",header=TRUE,sep=" "): Reads the space-separated text file named data.txt in the working directory into a data frame named Mydata. The first line of data.txt is a header containing variable names.

Mydata=read.table("data.txt",header=FALSE,col.names=c ("length","width","height","weight"),sep=","): Reads the comma-separated text file named data.txt in the working directory into a data frame named Mydata. Data begin in the first line of `data.txt` without a header. Variables are given names in the text vector `col.names`.

`attach(Mydata)`: Attach the data frame `Mydata` (makes the variable names available in R as vectors).

`Mydata=data.frame(length,width,height,weight)`: Combines vectors `length`, `width`, `height`, `weight` into a data frame named `Mydata`.

`write.table(Mydata,file="data.txt",sep=" ")`: Writes the data frame named `Mydata` into a space-separated text file in the working directory named `data.txt`.

`detach(Mydata)`: Detach the data frame named `Mydata` (removes the variables as vectors).

`x=scan("data.txt",sep=" ")`: Reads data in the file data.txt in the working directory into the vector `x`, line by line. Data in the file are numeric and space separated.

`A=matrix(scan("data.txt"),byrow=TRUE,ncol=5,sep=" ")`: Matrix A with five columns constructed by reading elements row by row from the space-separated text file `data.txt` in the working directory, which must contain all numeric data.

`write(A,"data.txt",sep=" ")`: Writes the matrix A into a space-separated data file named `data.txt` in the working directory.

`A=data.matrix(Mydata)`: Convert the data frame named `Mydata` into a matrix named A. Categorical and logical variables are converted into numeric variables.

Loops and Conditional Execution

```
for (i in a) {
  statement 1
  statement 2
  ⋮
}
```

The statements *statement 1*, *statement 2*, ... are executed repeatedly while i takes each value in vector a.

```
while (logical statement) {
  statement 1
  statement 2
  ⋮
}
```

The statements *statement 1*, *statement 2*, ... are executed repeatedly as long as *logical statement* remains true.

```
if (logical statement) {
   statement 1a
   statement 1b
   ⋮
} else {
   statement 2a
   statement 2b
   ⋮

}
```

The statements *statement 1a*, *statement 1b*, ... are executed if *logical statement* remains true. Otherwise, *statement 2a*, *statement 2b*, ... are executed.

Matrices

A=matrix(x,m,n): Vector x is read into an m by n matrix A, row by row. Values in x are recycled.

mat.or.vec(m,n): Creates an m by n matrix of zeros. Same as matrix(0,m,n).

rbind(a,b,c,...), cbind(a,b,c,...): Row bind, column bind. Concatenate vectors a, b, c, ... together as rows or columns of a matrix.

A+B: Matrix addition.

A-B: Matrix subtraction.

A%*%B: Matrix multiplication.

A^n: Elementwise power (each element of A raised to power n).

t(A): Transpose (turn rows into columns) of matrix A.

diag(A): Extracts main diagonal of matrix A as a vector (argument A is a matrix).

diag(n): Constructs identity matrix with n rows (argument n is a positive integer).

diag(x,m,n): Constructs an m by n matrix with diagonal elements given by vector x when going from upper left to lower right in the matrix, with all other matrix elements set to zero.

rowSums(A), colSums(A): Sums the rows or columns of matrix A and returns a vector.

rowMeans(A), colMeans(A): Vector of arithmetic means of the rows or columns of matrix A.

det(A): Determinant of the square matrix A. The determinant is a measure of the generalized (and signed) volume of the space contained by the columns (or rows) of A. If the determinant of A is zero, then inverse of A does not exist.

solve(A): Inverse of the square matrix A, if it exists.

solve(A,b): Solution to the system of linear equations given by Ax=b.

eigen(A): Eigenvalues and eigenvectors of the square matrix **A**. An eigenvector is a vector **v** for which multiplying the vector by the matrix **A** is just like multiplying the vector by a constant λ: $\mathbf{Av} = \lambda\mathbf{v}$. If **A** is $n \times n$ and the inverse of **A** exists, there are n different eigenvalues $\lambda_1, \lambda_2, \ldots, \lambda_n$ and corresponding eigenvectors $\mathbf{v}_1, \mathbf{v}_2, \ldots, \mathbf{v}_n$ of **A**.

Graphics

Plots of One Variable

stripchart(x): Stripchart (or Cleveland dot chart) of the elements in vector x.

hist(x): Histogram of the elements in vector x. Optional argument breaks=a provides bounds (elements of vector a) to the histogram groups. Optional argument freq=FALSE produces a relative frequency histogram rather than the default frequency histogram.

stem(x): Stem-and-leaf plot of the elements in vector x.

boxplot(x): Boxplot of the elements in vector x.

plot(x): Plot of elements in vector x (vertical axis) versus the index vector 1:length(x) (horizontal axis).

barplot(x): Bar graph of the elements in vector x.

pie(x): Pie chart of the elements in x.

Plots of Two Variables

plot(x,y): The x–y plot of vectors x and y, with x on the horizontal axis and y on the vertical axis.

matplot(x,Y): The x–y plot of every column of matrix Y versus vector x, on the same graph.

matplot(X,Y): The *x–y* plot of every column of matrix Y versus corresponding column of matrix X, on the same graph.

pairs(X): X is a matrix or data frame: Scatterplots of all pairs of columns (SPLOM).

Plots of Three Variables

persp(x,y,Z): Wire mesh (surface) plot of matrix Z (vertical axis) versus vectors x and y.

contour(x,y,Z): Contour plot of matrix Z on *x*- and *y*-axes. Optional argument nlevels= sets the number of contour levels.

plot3d(x,y,z): Three-dimensional scatterplot (spin with mouse) of vectors x, y, z. Must first load the rgl package with the command library(rgl).

Multiple-Panel Graphs

layout(matrix(c(1,2,3,4,5,6),3,2)): Creates a figure with six panels: Three rows and two columns. Plots will be added to the figure as they are created, in the order of the numbers in the matrix, panels 1 and 2 in row 1, etc.

Optional Arguments in Most Plotting Functions

pch=: Plotting characters (0: square, 1: circle, 2: triangle, 3: cross, 4: X, 5: diamond, 6: upside-down triangle, and other symbols shown in Chapter 13, Figure 13.1).

lty=: Line types (0: no line, 1: solid line, 2: dashed line, 3: dotted line, 4: dash-dot line, 5: long dash line, and 6: long dash–short dash line; see Chapter 13, Figure 13.2).

type=" ": Type of plot.

p: Only symbols for points drawn; no connecting lines

l: Only connecting lines drawn; no symbols for points

o: Both symbols for points and connecting lines drawn

b: Same as o except the connecting lines do not touch the point symbols

h: Vertical line from each point to the horizontal axis (comb plot)

s: "Staircase" plot, with neighboring points connected by a horizontal line followed by a vertical line

xlim=, ylim=: Limits to *x* and *y* axes. Example: xlim=c(0,10).

xlab=" ", ylab=" ": Text labels for *x* and *y* axes.

`xaxt="n"`, `yaxt="n"`: Supresses drawing of *x* and *y* axes.

`main=" "`: Plot title.

`sub=" "`: Plot subtitle.

`lab=`: Number of tick marks on *x* and *y* axes. Example: `lab=c(7,3)` gives an *x* axis divided into eight intervals by seven tick marks and a *y* axis divided into four intervals by three tick marks.

`tcl=`: Length (fraction of text height) and orientation (positive: in, negative: out) of tick marks.

`axes=FALSE`: Suppresses drawing of axes.

`ann=FALSE`: Suppresses drawing of axis labels and titles.

`cex=` (and `cex.axis=`, `cex.lab=`, `cex.main=`, `cex.sub=`): Size multiplier of plotting characters, axis widths, axis labels, title text, subtitle text.

`lwd=`: Line width multiplier.

`col=` (and `col.axis=`, `col.lab=`, `col.main=`, `col.sub=`): Colors of plotting symbols and lines, axes, axis labels, titles, subtitles.

Global Graphics Options

`par()`: Set graphical options globally for ensuing plots. Most options can also be set locally as arguments in plotting functions for individual plots.

`pin=c(width,height)`: Set plot dimensions in inches (only available as an option in the `par()` function).

Lines, Text, Titles, and Legends: Modifications to Existing Graphs

`points()`, `matpoints()`: Adds points to plots made with `plot()` and `matplot()` functions, with arguments similar to `plot()` and `matplot()`.

`lines(x,y)`, `matlines(x,y)`: Connects points given by *x* and *y* coordinates in vectors *x* and *y* with lines on plots made with `plot()` and `matplot()` functions.

`segments(x0,y0,x1,y1)`: Adds line segments to plots. Segments originate from *x* and *y* coordinates in vectors x0 and y0 and terminate at coordinates in x1 and y1.

`abline()`: Adds a horizontal line, vertical line, or line with designated intercept and slope to a plot.

abline(a,b): The intercept is a and b is the slope of a single line to be drawn.

abline(h=): Draws a horizontal line at the *y* value specified by h=.

abline(v=): Draws a vertical line at the *x* value specified by v=.

arrows(x0,y0,x1,y1): Draws arrows with origins at coordinates in vectors x0, y0 to coordinates in vectors x1, y1. Specify length (in inches) and angle (in degrees) of arrowhead edges in length= and angle= arguments.

text(x,y,c("a","b","c"...)): Adds text elements in vector c("a","b","c"...) to a plot at coordinates given by elements of vectors x and y.

title(main=c("line 1","line 2",...)): Adds a title with multiple lines to a plot.

title(main="bigger",sub="smaller"): Adds a title and subtitle to a plot.

legend(): Adds a legend to a plot.

Example:

legend(3,1,c("sine","cosine"),lty=c(1,2)): Adds legend box with upper left corner at plot coordinate (3,1), with text labels "sine" and "cosine," and line types 1 and 2.

windows(),quartz(),X11(): Opens a new graphics window (Windows, Mac OS, Unix/Linux).

Probability and Statistics

Statistics

summary(Mydata): Summary statistics for variables in the data frame Mydata.

aggregate(x,by=list(g1,g2,g3),FUN=*function*): Applies the specified *function* for summary statistics (such as mean) to the vector x separately within groups designated by categorical variables g1, g2, g3, ...

t.test(y): One-sample t-test of the null hypothesis that mean of variable y is zero.

wilcox.test(y): One-sample nonparametric Wilcoxon test of the null hypothesis that mean of variable y is zero.

t.test(y~group): Two-sample t-test of the null hypothesis that the two means corresponding to the two levels of a categorical variable group are equal.

`wilcox.test(y~group)`: Two-sample nonparametric Wilcoxon test of the null hypothesis that the two means corresponding to the two levels of a categorical variable `group` are equal.

`results=lm(y~group)`: One-way analysis of variance test of equality of means corresponding to three or more levels of a categorical variable `group`. Print the table with `summary(results)`.

`results=lm(y~group+trt)`: Analysis of variance of factorial experiment with factors given by the categorical variables `group` and `trt`, additive effects only (no interaction). Print the table with `summary(results)`.

`results=lm(y~group*trt)`: Analysis of variance of factorial experiment with factors given by the categorical variables `group` and `trt`, including interaction. Print the table with `summary(results)`.

`results=lm(y~x)`: Linear regression with y as response variable and x as predictor variable. Print the table with `summary(results)`.

`results=lm(y~x1+x2+x3)`: Multiple regression with y as response variable and x1, x2, x3 as predictor variables. Print the table with `summary(results)`.

`results=lm(y~x+group+x*group)`: General linear model with y as response variable, x as quantitative predictor, and `group` as categorical predictor. Print the table with `summary(results)`.

`results=nls(y~b1*x/(b2+x),data=Enzyme,start=list(b1=b1.0, b2=b2.0))`: Nonlinear least-squares fit of enzyme rate equation to enzyme rates in variable y with substrate concentrations in variable x, where y and x are in the data frame `Enzyme` and initial values of constants are given in the `start=`argument. Print the table with `summary(results)`.

Sampling

`sample(x,size,replace=TRUE),sample(x,size,replace=FALSE)`: Draw random sample from vector x of size given by `size`, with or without replacement.

Probability Distributions

Binomial, n trials, probability of success p:

  `dbinom(x,n,p)`: Probability of x

  `pbinom(x,n,p)`: Summed left tail probabilities up to and including x

  `qbinom(q,n,p)`: The qth quantile ($0 < q < 1$)

  `rbinom(m,n,p)`: Vector of m random numbers

Poisson, mean `lambda`:

  `dpois(x,lambda)`: Probability of x

`ppois(x,lambda)`: Summed left tail probabilities up to and including x

`qnbois(q,lambda)`: The qth quantile (0 < q < 1)

`rnbois(m,lambda)`: Vector of m random numbers

Negative binomial, shape k, probability p:

`dnbinom(x,k,p)`: Probability of x

`pnbinom(x,k,p)`: Summed left tail probabilities up to and including x

`qnbinom(q,k,p)`: The qth quantile (0 < q < 1)

`rnbinom(m,k,p)`: Vector of m random numbers

Multinomial, number of trials n, vector of probabilities p:

`dmultinom(x,n,p)`: Probability of vector x

`rmultinom(m,n,p)`: m random vectors

Uniform on interval (0,1) or (a,b):

`dunif(x)`, `dunif(x,a,b)`: Density at x

`punif(x)`, `punif(x,a,b)`: Cumulative probability at x

`qunif(q)`, `qunif(q,a,b)`: The qth quantile (0 < q < 1)

`runif(m)`, `runif(m,a,b)`: Vector of m random numbers.

Exponential, `beta=1/mean`:

`dexp(x,beta)`: Density at x

`pexp(x,beta)`: Cumulative probability at x

`qexp(q,beta)`: The qth quantile (0 < q < 1)

`rexp(m,beta)`: m random numbers

Gamma, shape=k, rate=beta (mean=k/beta):

`dgamma(x,k,beta)`: Density at x

`pgamma(x,k,beta)`: Cumulative probability at x

`qgamma(q,k,beta)`: The qth quantile (0 < q < 1)

`rgamma(m,k,beta)`: m random numbers

Chi-square, degrees of freedom=df:

`dchisq(x,df)`: Density at x

`pchisq(x,df)`: Cumulative probability at x

`qchisq(q,df)`: The qth quantile (0 < q < 1)

`rchisq(m,df)`: m random numbers

Normal, mean=mu, standard deviation=sigma (omitting the mu, sigma arguments produces normal with mean 0 and standard deviation 1):

`dnorm(x,mu,sigma)`: Density at x

`pnorm(x,mu,sigma)`: Cumulative probability at x

`qnorm(q,mu,sigma)`: The qth quantile ($0 < q < 1$)

`rnorm(m,mu,sigma)`: m random numbers

Student's t distribution, degrees of freedom=`df`:

`dt(x,df)`: Density at x

`pt(x,df)`: Cumulative probability at x

`qt(q,df)`: The qth quantile ($0 < q < 1$)

`rt(m,df)`: m random numbers

F distribution, numerator degrees of freedom=`df1`, denominator degrees of freedom=`df2`:

`df(x,df1,df2)`: Density at x

`pf(x,df1,df2)`: Cumulative probability at x

`qf(q,df1,df2)`: The qth quantile ($0 < q < 1$)

`rf(m,df1,df2)`: m random numbers

Getting Help

`help()`: Get help with a known function. Example: `help(plot)`.

`??string`: Search for a topic (text string) in documentation.

Example:

`??t-test.`

Index